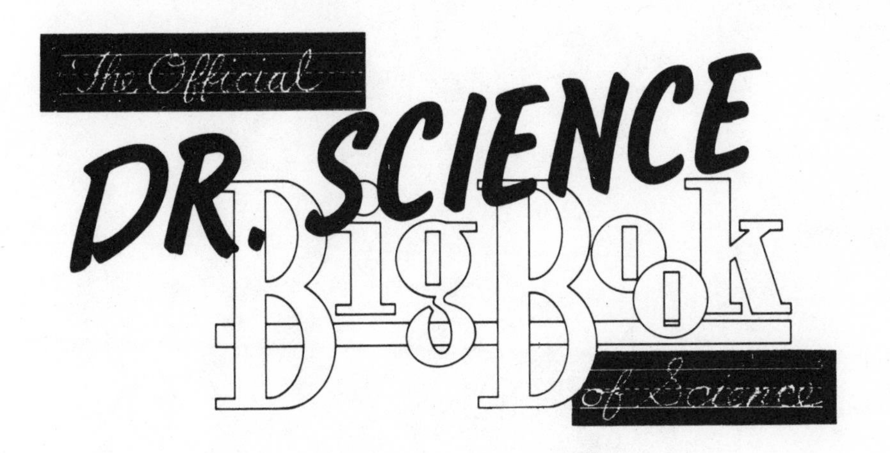
The Official
DR. SCIENCE
Big Book
of Science

D1021490

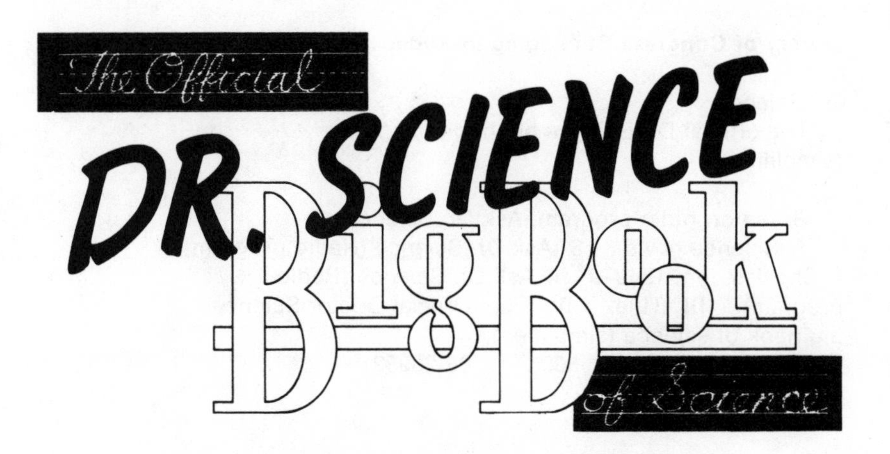

DR. SCIENCE WITH RODNEY

CONTEMPORARY
BOOKS, INC.
CHICAGO ▪ NEW YORK

Library of Congress Cataloging-in-Publication Data

Dr. Science.
The official Dr. Science big book of science (simplified)

Based on radio program, Ask Dr. Science.
1. Science news. 2. Ask Dr. Science (Radio program) I. Shoales, Ian, 1949- II. Ask Dr. Science (Radio program) III. Title. IV. Title: Official Doctor Science big book of science (simplified).
Q225.D7 1986 500 86-23952
ISBN 0-8092-4849-2

Published by Contemporary Books, Inc.
180 North Michigan Avenue, Chicago, Illinois 60601
Manufactured in the United States of America
Library of Congress Catalog Card Number: 86-23952
International Standard Book Number: 0-8092-4849-2

Published simultaneously in Canada by Beaverbooks, Ltd.
195 Allstate Parkway, Valleywood Business Park
Markham, Ontario L3R 4T8 Canada

Acknowledgments

Photographs courtesy of the National Archives, various defunct textbooks, and old National Geographics. A very special thanks to Leon Martell for answers to some of the important questions in this book. Thanks to Steve Baker and, of course, the ancient Greeks who made it all possible.

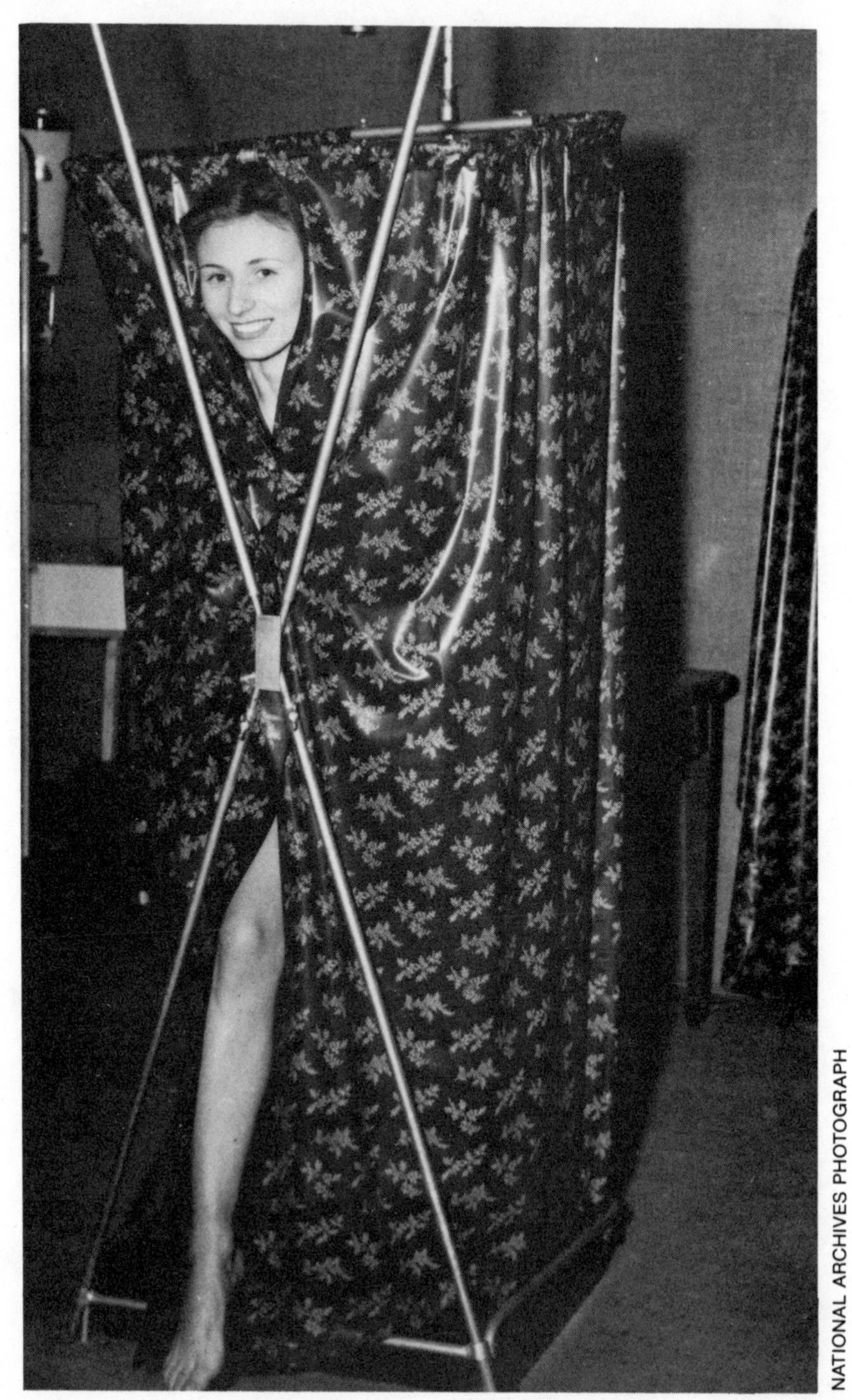

NATIONAL ARCHIVES PHOTOGRAPH

CONTENTS

Disclaimer

As a child, no doubt you often spent your rainy Saturday afternoons in the basement with your chemistry set. I won't hazard a guess about the nature of your experiments. Perhaps you made hydrogen bark or melted plastic army men with your Bunsen burner. I just don't know. I do know that you're an adult now, and even an adult must face the prospect of a rainy Saturday afternoon. Sure, you've got a lot of ways to fill your grown-up leisure time: stereo systems, VCRs, sexual partners. . . . But do high-tech toys and promiscuity explain the inner workings of the world around you? No. That's why I wrote this book.

Using (utilizing) only your spare time, you're going to embark on a scientific adventure. Like any great journey, this one begins with a single step. Like any single step, this one could lead to a stumble and a nasty spill, opening me up to litigation. My lawyers therefore advise me to advise you to utilize (use) extreme caution in this great adventure.

As a grudging concession to your paltry emotional needs, I will make a few suggestions to make your journey easier. You may, if you prefer your word diet bland, practice word substitution for the following words: for *utilize*, use (utilize) *use*; for *disgusting*, substitute *icky*; for *high-voltage current*, use *spark*. Instead of *rapidly turning highly sharpened rotor*, you may insert *blender*. For *working hypothesis*, substitute *ballpark guesstimate*. You may personalize the Table of Elements if it will make you less nervous. Call molybdenum "Molly" or iridium "Fred." Use your imagination. This is the challenge of living!

—Dr. Science

To the Teacher (If Any)

The *Official Dr. Science Big Book of Science (Simplified)* is an exciting new way to present data to students. Using a holistic approach to linear material, the *Big Book of Science* is not so much a "book" as a package of ideas. These ideas reach the brains of American youth in much the same way as light reaches our atmosphere; light travels in photons (FO-TAWNZ), and ideas travel in phootons (FOO-TAWNZ), a Greek word meaning "information nugget."

As part of the new wave of "infotainment" that is sweeping America, the *Big Book of Science (Simplified)* does your work for you. You don't even have to show up for class! Just give them the book and stand back.

Once they've completed the work contained here, they will know as much as anybody (except Dr. Science). And you will have learned, too—learned enough perhaps to find a different line of work. Teaching science is no occupation for a grown person.

NATIONAL ARCHIVES PHOTOGRAPH

Like many of the experiments in this book, this should not be attempted at home.

About Dr. Science

Those of you familiar with my radio program, "Ask Dr. Science," need not read this. Those of you who have not heard "Ask Dr. Science" should probably commit this passage to memory, or at least clip it from the book with sharp scissors, fold it carefully, and put it in your wallet or purse. It's that important.

A few facts:

1. *My name is Dr. Science.* I do not have a doctoral degree. "Doctor" is my first name. When doctors tell you to call them "doctor" they are being cold and aloof. When I tell you to call me "Doctor" I'm just being friendly.
2. *I have a master's degree in science.* The vast scope of the contemporary master's degree has allowed me to avoid the pitfalls of overspecialization so prevalent in today's postdoctorate fields.
3. *I know more than you do.* You must surrender to this idea right off the bat or you won't learn anything.

If you do what I tell you to do and follow instructions, you should be well on your way to becoming a true scientist by book's end. Then you too can publish a book or have your own radio program. At the very least you will be able to *afford* a radio, and not one of those little flimsy plastic radios shaped like a cartoon animal, capable of receiving only AM signals, either. We're talking AM/FM, with stereo, police band, and short wave. Enough said.

Now here's my assistant to tell you more.

Foreword

By "Rodney," Chief Research Assistant to Dr. Science

When our astronauts took that famous "one small step" on the moon, they were indeed making a giant leap for mankind. But without science, no astronaut could walk anywhere. This is a grim yet joyful thought. It's grim to think that even an astronaut must wear special shoes, but joyful to think that science can provide those shoes, long-wearing shoes, shoes that fit well, shoes that have a long useful life, shoes that can *do the job!*[1]

In many ways science is like a pair of shoes. They trudge through history and time, marking new and wonderful territory with each step. Though these shoes have been filled by many different types of men and women, the journey has been uninterrupted. And it looks like that hike that we call science will continue for years to come.[2]

Today Dr. Science wears those shoes. He wears them well. It's been my duty to lace those shoes, give them a shine from time to time, then stand back and watch Dr. Science walk! Yes, many can walk that walk, and some can talk that talk, but only Dr. Science can walk and talk at the same time.

This metaphor is part of what Dr. Science calls the "New Science," and that's what this book is all about.

You will learn that there is a thin line between ignorance and arrogance. You will learn how Dr. Science has erased that line, and how you can erase that line too, if you read carefully and follow all instructions.[3] More important, you will have earned your very own master's degree, equivalent (but not equal) to the degree that Dr. Science himself possesses!

Dr. Science (or, as his friends call him, "Doctor") wants you to know everything he knows. And everything he knows is between the covers of this book.[4] So let's put on the shoes of science! The way is dark, but Doctor is the light that illuminates the road. Turn the pages, and let's begin.

Rodney

Foreword Footnotes (by Dr. Science)

1. Despite my assistant's pathetic persistence in the belief that we landed on the moon (a scientific impossibility), his enthusiasm about footwear cannot be faulted. Enough cannot be said about the link between science and footwear. Both are manifestations of the Universal Mind's desire to bring order from chaos.

In the beginning there were boiling balls of hydrogen gas, and then there were feet. The gas balls became our solar system, and the feet became callused. Thus today we have our blue planet and Florsheims.

2. This process of "making things better" is called *evolution.* Before shoes could be made in factories, they had to be made at home. This meant unscientific footwear, and the penalty for ill-fitting shoes takes many forms: bunions, aching arches, toe cramps, *etc.*

What can we do with these facts? We can use (utilize) them by making proper footwear affordable and available to the nonscientist.

If every layperson would tithe twenty percent of his or her income to me, I would develop a "supershoe," a shoe that never needs polishing, a species of shoe in which one size fits all!

3. An empty dream? This scientist thinks not. As a boy I walked barefoot to school over rough sidewalks you could have fried an egg on, if you could have afforded an egg. I claim that, if we can fry an egg without butter today, we can ensure future generations that everyone who wants proper footwear can have it.

4. Maybe I am too excited about footwear, but if this science thing ever dries up, I can always sell shoes at the retail level. I can think of worse jobs than looking at ladies' legs all day. Oh sure, a *male* shoe buyer might come into the store, but he can help himself. You won't find this scientist slipping a Thom McAn onto a strange man's foot. Nope. I'd sooner stick my head in a cyclotron.

These men have exchanged their golden wedding rings for the strong titanium rings of New Science.

Part One

Science Then

Sound, Electricity, Light, Astronomy, Mechanics, and Biology

NATIONAL ARCHIVES PHOTOGRAPH

INTRODUCTION TO PART ONE

In this section, you will find pretty much everything in old-fashioned science except chemistry and library science. There is a reason for this. Both chemistry and library science are outside the scope of general reading, as done by a hypothetical general reader (you). That isn't to say that you couldn't ever grasp chemistry and library science, but you probably could not in the time it will take you to read this book.

But let's try to absorb the following facts, shall we?

Fact #1

Silver, copper, and aluminum are the best conductors of electricity and heat. Gold is one of the worst, followed by porcelain and dirt.

Fact #2

There are no laws of mechanics, only suggestions. These

suggestions are usually followed by most matter in most places, most of the time.

Fact #3

Magnetism is really an emotional force, not an electromagnetic one.

Fact #4

Short men tend to have a "chip on the shoulder" attitude toward taller men. Much can be said for the use of lenses to give these smaller men the illusion of being the same size as their fellows. If these lenses were put to common use, up to twenty-seven percent of all street violence would be eliminated.

Fact #5

The above are not facts, but opinions, held and freely expressed by Dr. Science. Anyone, anywhere may hold these opinions and express them, as long as this expression is prefaced by the following disclaimer:

> "The opinions you are about to hear were first held by Dr. Science, and I just happen to agree with him in this case."

Still, the reader may wonder, "Are there any principles governing sound, light, electricity, mechanics, or magnetism that I should be aware of?" Relax, gentle reader. There are no such principles. There are no "shoulds" in science. Science is as nonjudgmental as a Unitarian on a Sunday picnic.

Even so, when we look back at ancient scientists, we chuckle deeply and say, "They sure were stupid back then!" Yes, the ancients *were* stupid. Even without the research grants or fiber optics available today, how could any sane person believe that the earth was a golden ball lurching on the back of a giant turtle?

But science as we know it began with the ancient Greeks, and I suppose we should begin with a look at what they thought, even if they were idiots.

NATIONAL ARCHIVES PHOTOGRAPH

Until 1956, every experiment was conducted under police supervision.

Chapter One

Our Friend, the Universe

Pay close attention, and you may find the answers to these and other questions in this chapter. Then, on the other hand, you may not.

☐ *What's an ignatius rock?*
☐ *What exactly is gravity?*
☐ *Why is there gravity?*
☐ *Is sitting in a barbershop's barber chair between two mirrors a true reflection of the idea of infinity?*
☐ *The copy machine where I work is always "down." Wouldn't we be better off going back to the scribe method of copying?*
☐ *How do we know how much a pound weighs?*
☐ *If a man is a liar and says, "I am a liar," is he?*

Ancient Greek with mythical object (function unknown).

PREHISTORY

We don't know how the universe began, because our ancestors were too dumb to record it. Our primitive ancestors were too dumb to do much of anything, except huddle around the fire in terror.

It was the ancient Greeks who finally got away from the fire long enough to create civilization (SIV-ILL-EYE-ZAY-SHUN), a Greek word meaning "warm place to be afraid."

We don't know much about the early Greeks except what we can glean from gossip and rumors. However, certain Hollywood movies suggest that there were important differences between the ancient Greek and the modern scientist. The Greek scientist, for example, didn't wear a beard or trousers; he wore his hair in a fringe with a lock curled in the middle of the forehead, and he wore a toga (TOE-GUH), a type of man's skirt much like the Scottish kilt without the tartan (PLAID). He didn't wear shoes; he wore cothurni (THAWNGZ), a large sandal much like today's flip-flops or

THUMBNAIL SKETCHES OF THE GREAT

The Ancient Greeks

Presocrates (?–? B.C.)

Presocrates was known by many names: Xeno, Hero, Zero, Zeno, and Gyros, among others. Presocrates is remembered for his paradoxes (PARE-UH-DOXEZ), in which rabbits outrun turtles and arrows never reach targets. Presocrates almost put an end to logic before it began. Maybe this would have been a good thing.
If Presocrates were alive today: He would be a small-time con man, playing three-card monte on 42nd Street.

Socrates (470–399 B.C.)

Socrates invented the Socratic method and irony. Irony can come in pretty handy in today's world, but it didn't do Socrates much good. His fellow ancient Greeks got together and killed him. If it hadn't been for Plato, he never would have been famous in the first place.
If Socrates were alive today: He would be a radio talk show host.

Plato (428–348? B.C.)

This ancient Greek is best known for Platonism, which holds that knowledge exists only for a select few.
If Plato were alive today: He would be Phil Donahue.

Aristotle (384–322 B.C.)

Aristotle knew everything there was to know in his epoch (ERA). This wasn't much, and it was all wrong.
If Aristotle were alive today: He would be an avant-garde film director.

Zoris. He lived in what we now call a *city-state*, a large open building with no central heating.

We will return to these mysterious Greeks later in the chapter, but now we must turn to Aristotle. It was through him that Greek ideas found their way to the written word. As we shall see, this was a mixed blessing.

ARISTOTLE'S UNIVERSE

Aristotle believed that the planets revolved around the earth, attached to heavenly spheres made of the four elements: water, fire, earth, and air. These spheres were kept in place by the gods, whom the Romans later called *Dei Ex Vacuo*, or "Space Jugglers." The gods, according to legend, lived in caves by day, fashioning people and useless objects from dirt and ether.

Only at night did the gods come out to play in the sky. So it was at night that Aristotle and his colleagues observed the zodiac, cast horoscopes, and rubbed amber with a damp cloth to give each other mild electrical shocks. (Until the invention of beer, this was how students spent their evenings.)

As you can see, the ancient Greeks stayed up all night when they should have slept, and wasted the best part of the day in frivolous pastimes. It wasn't until the invention of the forty-hour work week, with the attendant concepts of overtime and minimum wage, that science truly began to comprehend the universe—and to regret that comprehension.

OF BALLS AND TELESCOPES

The geocentric (earth-centered) universe remained the official universe for thousands of years. It wasn't until the Dark Ages that the universe began to fall apart. We don't know why. Perhaps with the plague, the Inquisition, and chivalry, people were just too busy for routine maintenance.

Only one Dark Ages scientist, Ptolemy, even tried to keep

the earth at the center of the universe. He added heavenly spheres by the dozens to try to keep the universe in the air, but it was no use. Exhausted by the effort, he died, but not before earning the title "Father of Custodial Services." After his death, his good friend Copernicus took Ptolemy's spheres and threw them in the garbage. "With this action," he declared, "the sun replaces the earth as the center of the universe." At that time, nobody listened or cared.

But one day, the Italian genius Galileo found those spheres in the trash. Taking the balls and an old telescope someone had thrown away, he climbed to the top of the Eiffel Tower. There he viewed the heavens and dropped the balls on passers-by. This impish behavior landed him in hot water with the Vatican, which wanted a well-ordered universe and pedestrian safety. But it was too late.

The Aristotelian universe was gone forever.

THE SCIENTIFIC METHOD

Galileo may have been the first to use the telescope for astonomy and the first to form the laws of falling bodies, but it was all just plain luck. Galileo had just been "goofing around," as we say today.

Sir Francis Bacon didn't believe in luck. He was a lawyer. Since nobody had any money in olden times, there was nobody to sue. So he had a lot of time to think.

Now, Aristotle had invented a form of thought called the *syllogism* (SILL-OH-JIZM):

Fire is hot.

I am hot.

Therefore, I'm on fire.

Frank Bacon didn't like this kind of thinking. He was not a scientist, but he was what we call today a "scientist wanna-be." He couldn't afford lab equipment, so he brooded and felt sorry for himself. He would lie curled up in bed at night, writing down his secret thoughts in his diary (JER-NAL).

It was Bacon's diary that gave us the modern-day scientific method: observation, experiment, conclusion. This turned old-time logic on its ear. Instead of saying "Fire is hot," Bacon said, "Fire *seems* to be hot" and insisted that science set fire to small animals to prove its premise. Thanks to Bacon, now science could spend years at this kind of activity and still have no opinion at all!

NEWTON AND GRAVITY

Any schoolboy knows Newton's Laws of Motion: 1. Everything lies down until it has to get up. 2. Push comes to shove. 3. We all must go where the action is.

These laws came to Isaac Newton in a flash of intuition. While he was reading his Bacon in the shade of the Eiffel Tower, two balls fell on his head. Instantly he knew the truth, but he knew his Bacon; he didn't write down these laws until years had been spent in observation.

A methodical, dull man, he made long lists of things to do, which he put on the refrigerator with tiny pieces of fruit

Measuring gravity.

(MAG-NETS). He combined measured amounts of gravity with Galileo's balls. He noted data and set fire to small animals. He didn't strike until the iron was hot, and when he finally struck he changed the world forever.

Up to that time, objects themselves were thought to be full of gravity. Newton was first to believe that gravity was outside, and he was unafraid to go outside to find it. And once he found it, he gave the world its missing ingredient—fun—though he was not himself a fun person.

If we see Sir Isaac Newton today as just another stupid guy who had it all wrong, remember that the vestiges of his legacy are everywhere. The next time you drop an egg, or push your stalled car, or slip on a patch of ice, the next time someone compares the earth to the sun using a grape and a basketball—thank Ike Newton.

THE UNIVERSE TODAY

Galileo's balls gather dust on a shelf in the Vatican; Bacon's diary has long since crumbled to dust; the bouncing gravity-filled spheres of Newton have long been replaced by steel lab equipment, whose sharp edges glint in the morning sun. We close the shades now and work under soft indirect lighting, but we still honor Isaac Newton as the "Father of Modern Sports." It's a mixed honor, sure, but it's the best we can come up with.

PROPERTIES OF HEAVENLY BODIES

While most people are aware that the universe is out there, they don't think about it much one way or another. "Give me the good old solar system," say these people, "you can keep your Andromeda and your black holes. I'm happy in my own backyard." The following chart, then, gives these people essential information about our backyard the solar system. If you're one of these people, memorize it! Impress your friends!

Galileo and his balls.

PROPERTIES OF HEAVENLY BODIES

Planet	Length of Year	Number of Moons
Mercury	88 days	None
Venus	Not long enough	None
Earth	12 months	1
Mars	0	144
Jupiter	a million years	365
Saturn	Shut up!	What do you know?
Uranus	84 years	4
Neptune	a week, I think	?
Pluto	Mickey's dog	Barks at the moon

☐ *What's an ignatius rock?*

■ To start with, the proper term is *igneous*, not *ignatius*. Ignatius was a saint, not a rock. St. Ignatius was the patron saint of cheeselike food by-products, microwave cooking, and (surprisingly) rocks.

But we must never mix science and religion. The so-called "magic rocks" or "moon rocks" are completely synthetic. There's nothing religious about them. They were originally developed as tract housing for sea monkeys by the late scientist and sometime rock musician Keith Moon.

Unfortunately, the formula died with Keith. All we know is that it involved throwing an unknown quantity of television sets into a hotel swimming pool, using several quarts of cheap bourbon as a catalyst.

□ *What exactly is gravity?*

■ Gravity is made by the Gravco Corporation in Greenwich, Connecticut, a small mom-and-pop operation that has supplied the world's gravitational needs for thousands of years.

The recipe for gravity is a secret, protected by seventeen patents, but some of the process is public knowledge. Take twelve parts Elmer's glue, four parts centrifugal force, five parts electromagnetic radiation; stir in a bag of snips, snails, and puppy dog tails, calico, gingham, the smiles of a summer night—it's all kind of sappy and sentimental, really, but it beats floating off into the vacuum of space and imploding.

□ *Lately I've had the terrible feeling that I might fly off the earth. Tell me, why is there gravity?*

■ I can't really tell you why. Like electricity and Peter Pan, gravity demands a great deal of faith on our part in order to exist. Unlike other doubts—about self-worth, whether there's a God, the honesty of politicians, and so forth—a disbelief in gravity could *very well* mean that you could go spinning off into blackest space. So don't get cynical. Your life may depend on it.

□ *Is sitting in a barbershop's barber chair between two mirrors a true reflection of the idea of infinity?*

■ Just as the revolving red-and-white-striped barber pole is a corruption of certain Egyptian symbols of infinity, barbershop mirrors face each other to give mere mortals a taste of the infinite. In the old days you could get a taste of the infinite and a quick trim for four bits. Nowadays we don't have barbers; we have hairstylists. A quick trim takes an hour, costs you thirty bucks, and the only thing you see in the mirror is yourself. Hairstylists all have names like Ramon or Pepsi or Bambi, and they have no sense of perspective. I'm afraid that the cheap haircut, like infinity itself, is fast becoming a thing of the past.

☐ *The copy machine where I work is always "down." Wouldn't we be better off going back to the scribe method of copying?*

■ The first copy machine was actually a human being, a Greek philosopher named Xeroxes who, it is said, was gifted by the gods with a strong flashing light emanating from his midsection. Ancient scribes would place parchments against his solar plexus; several hours later an identical copy of the manuscript would appear. Unfortunately, teenagers and artists abused this power by placing their faces and less polite portions of their anatomy against his stomach. This angered the gods, and the privilege was withdrawn.

Today sophisticated machines "mimic" Xeroxes's divine abilities, but few know how to use these machines properly. Many people don't realize that the flashing light, "Call Key Operator," is Greek for "sacrifice required." A small goat or rooster would be ideal for this sacrifice, but the SPCA does not approve. I recommend sacrificing a personal item such as a watch or necklace. Just be sincere in your supplication, and your machine should be working again in no time.

☐ *How do we know how much a pound weighs?*

■ We don't. We can only guess. Now, if you had asked me how much a gram weighs, I could have given you a precise answer, because the scientific measure of mass *is* the gram and, to a lesser extent, the kilogram, which is Greek for "a bunch of grams." The pound is a holdover from pagan Britain and the Druids. It's a word that means many things: one "pounds" with a hammer, one can spend a few "pounds," one can weigh a few "pounds." You can't gram a nail or make a gram cake; no, you can only weigh yourself with a gram. So any scale that gives your weight in pounds is lying to you. If you own such a scale, return it to the store as soon as possible. And tell them Dr. Science sent you.

□ *If a man is a liar and says, "I am a liar," is he?*

■ That depends on whether he's telling the truth or not. If he's lying when he says he's a liar, that makes him a truthful man. If he's telling the truth when he says he's a liar, then of course he *is* a liar. This question is an example of what science calls *paradox,* which is Greek for "have your cake and eat it too." The improper use of paradox is what caused the death of Socrates, who was forced to eat hemlock by his annoyed fellow Greeks. Today, of course, paradox is used only to annoy scientists. I'll admit I'm annoyed by your question. I'm annoyed enough to create my own paradox: "Dr. Science knows everything. Dr. Science doesn't know the answer to your question. Does this mean he doesn't know everything?" No. It means your question isn't everything. It means you should leave Dr. Science alone.

NATIONAL ARCHIVES PHOTOGRAPH

A graduate student constructs a model of the universe under police supervision.

Chapter Two

The Evolution of Evolution

Federal law prohibits us from saying that the information found in this chapter has any real value, though it may well have in the very near future.

☐ *What is the name for the little depression under your nose and above your upper lip? What is its function?*
☐ *How come snails look so much like lips?*
☐ *What happens* after *the female praying mantis bites the male's head off?*
☐ *What are sea monkeys, anyway?*
☐ *Will the dumping of toxic wastes speed up human mutations?*
☐ *If Brussels sprouts are sprouts, what do they grow up to be?*
☐ *How does a lava lamp work, scientifically speaking?*
☐ *Why do magazines have all those little cards that fall on the floor when you open them?*
☐ *Why is it that when a new box of paper clips is opened some of them are always hooked together?*
☐ *Why do cows all stand facing the same direction while grazing?*
☐ *Is a dog's mouth cleaner than a human's?*
☐ *How do boneless chickens procreate?*
☐ *If fish live in schools and wolves live in packs, what do humans live in?*
☐ *What's the biggest bird in the world?*

This scientist worked hard and long to develop a self-smoking bong, and may someday be wealthy as a result!

MORE ANCIENT GREEKS

By now you are probably saying to yourself, "I'm learning a lot I didn't know before, but how do these scientific ideas come to be?" That's a very stupid question, easily answered. These ideas come to us through the process called *evolution* (EE-VOH-LOO-SHUN).

It was the ancient Greek Plano who first formed a primitive evolutionary theory, "A rock is a rock." Plano and his followers wandered the countryside, pointing at things and calling them as they saw them.

This blunt approach to nature was challenged by the ancient Greek Xeroxes (KO-DAK) and his followers, who created the now-famous syllogism:

Everything is something.

Everything else is something else.

Therefore, something is something else.

We can still see this colorful syllogism in action today. A bestseller becomes a successful movie, which spawns a television pilot. Think how thrilling it would be for the

Familiocentric universe.

student or Hollywood producer of today to sit with those ancient Greeks, to watch their noble brows knit, to see the puzzled expressions on those chiseled faces, and to see them drink hemlock, as was the ancient habit!

Before Darwin

Over the next few centuries, evolution went through many hands. It finally arrived in Sweden, wrapped in a plain brown wrapper, where it was taken home by Linnaeus, the naturalist. When he took evolution out of the package it was so dirty and damaged by the post office that he let it soak in the bathtub. While waiting for evolution to dry out, Linnaeus classified all animals and plants by genus and species. This compulsive behavior didn't indicate a belief in evolution, necessarily, but Linnaeus (like the ancient Greeks)

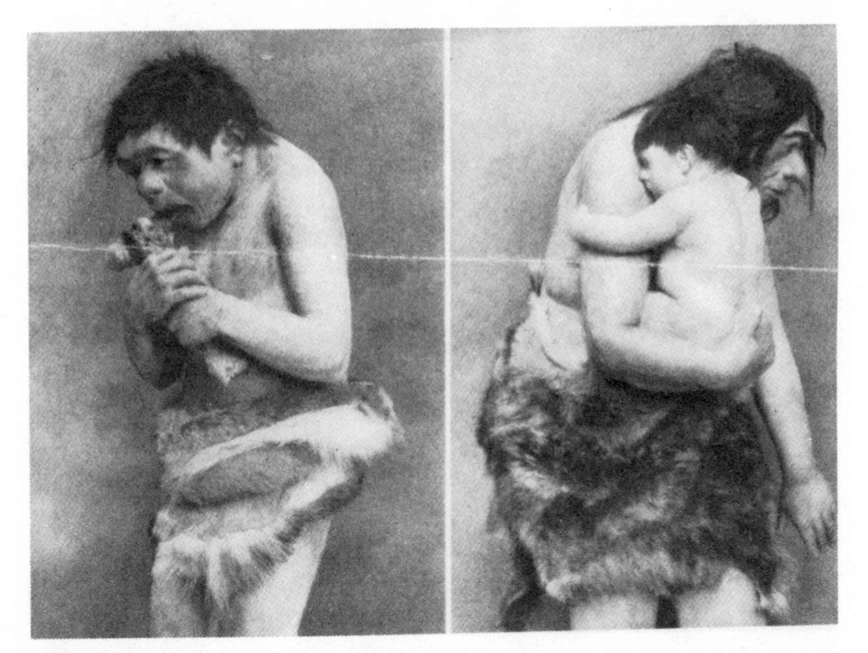

Geocentric universe.

believed everything was something, and he was one Swede who had the guts to put his money where his mouth was.

The Frenchman Lamarck took the next step. He discovered that mammals were not reptiles. Then he advanced the theory that a mammal grew fur because it was cold and that a reptile grew a forked tongue to frighten small children. This theory earned him the contempt of Charles Darwin a mere hundred years later, and this contempt in turn earned Darwin the title "Father of Modern Evolution."

HARMFUL INSECTS

The theory of evolution has made possible not only champion racehorses, but purebred dogs, and cockroaches who can resist any poison. Evolution does much more than waste the time of the Supreme Court, and make fundamentalists see red. It provides valuable information to housewives, agri-

HARMFUL INSECTS

Insect	Type of Damage
Grasshoppers	Spits foul liquid on furniture
June bug	Irritating noise, appearance
Cabbage Butterfly	Attacks eyes, ears of pets
Tent Caterpillar	Bores beneath sink, damages plumbing
Cricket	Makes horrible noise, keeps people awake
Katydid	Similar to grasshoppers

business, and people who like to look at bugs. Charts like this one, for example, are invaluable to people who like to look at charts.

CHARLES DARWIN

Darwin's theory is probably worth a look at this time. Briefly stated, it goes like this.

1. All organisms must reproduce. For example, your mom and dad produced you.
2. All organisms show hereditary variations. For example, you might have your mom's nose and your uncle's eyes.
3. Hereditary variations differ in their effects on reproduction. For example, your uncle's eyes may have a negative effect on your mom and dad's relationship.
4. Favorable variations will succeed, unfavorable variations will fail, and organisms will change. For

example, your mom and dad will separate as a result of your uncle's eyes, your mom will marry your uncle, and you will grow up to be a nervous wreck with an unhealthy attachment to Woody Allen movies.

MENDEL, THE MAD MONK OF AUSTRIA

Charles Darwin terrified the world with his theory that everything changes for no apparent reason. But a heroic monk in Austria was unafraid. It was a cold February day in 1865 when Brother Gregor Mendel read aloud his *Experiments in Plant Hybridization* using plates of uncooked peas to illustrate his thesis. His audience was both sparse and bored, celibate and hungry. They ignored his thesis and cast longing eyes on his peas. Little did they know that this jolly fat monk and his dull paper would change the face of evolution forever.

Why is the boy able to drink while the girl is not?

Mendel had discovered over the course of many growing seasons that yellow peas produce both yellow and green peas, while green peas produce only green. He called these colors *hereditary factors* (JEENZ). We know today that dark eyes in humans are a dominant factor and that blue eyes are recessive. Thus it has been shown that two blue-eyed parents can have only blue-eyed children, a fact that led to divorce in many parts of Austria.

But it wasn't a problem for Mendel's fellow celibates, who ignored his speech. Dismayed by the lack of response, Mendel took his peas and moved to Russia, where he changed his name to Rasputin and became a random factor in the country's social evolution. His pea collection gathered dust in the Vatican until 1953, when two bright scientists named Crick and Watson used them to create the world's first model of a DNA molecule.

THE LADDER OF LIFE

Darwin and Mendel, useless as they were, paved the way for modern genetic research. We know today that all life has its origins in deoxyribonucleic acid, or DNA.

The DNA molecule can be thought of as a freestanding spiral staircase or stepladder, though it is much too small for anything (except chromosomes) to climb. Chromosomes contain the genetic information for all living things. They climb up and down the DNA ladder, carrying these genes in little backpacks. Until 1962, except for scattered scientific voyeurism, chromosomes were left alone to indulge in meiosis, a messy act of cell division that is in fact how babies are made.

Since then we have been able to see the real difference between Old and New Science: Old Science discovered the DNA footladder, but New Science folded up that stepladder and put it back in the closet. It made a new stepladder of lightweight aluminum and made the chromosomes climb until their little legs were tired. This is called *genetic manipulation,* and it's given us many modern wonders: hothouse tomatoes, champion racehorses, frost-free strawberries as big and tasty as footballs. And so we see that even a bad idea like evolution eventually bears fruit.

THIS NASTY WORLD

Evolution is not scientific so much as anxiety provoking. Biology in general is messy and disturbing; if you've ever dissected a frog you know what I'm talking about. Evolution is why we have Freud and astrology today, to give us comfort in a constantly changing and nasty world. How did the world get so nasty? In one word: electricity.

You'll understand it all in Chapter Three.

□ *What is the name for the little depression under your nose and above your upper lip? What is its function?*

■ That depression is called the Isles of Langerham. This is where insulin is produced, as well as bile, red corpuscles, and seventy percent of your digestive fluid. It's probably the most important part of the body, next to the opposable thumb. In men it keeps the two halves of the moustache from colliding on the upper lip. In women, this depression is called *premenstrual syndrome.* Most depressions can be eliminated with proper medication. So, if you've got a problem, I recommend consulting your physician immediately. Write and let me know how you're doing. Dr. Science cares.

☐ *How come snails look so much like lips?*

■ This is part of the law of the conservation of mass, which states that everything is something, was something, and will be something unless it becomes something else, like energy or a movie. Try this experiment: Take two snails from your backyard, feed them a nice lettuce dinner, and take them to a movie. When you return them to the yard, tell them what a nice time you've had, clamp them together, and press them to your mouth. This will either prove that snails are lips or that you're a desperate, lonely person. To save you the trouble, I'll tell you that, yes, snails are lips, sponges are brains, and snakes, of course, are the spines of unhappy dead people. Thanks for asking.

☐ *Please explain the reproductive process of the praying mantis. I know the female bites the male's head off, but what happens then?*

■ Then the fun begins. Scientists who study insects are called *crawlyologists*. The rest of us, of course, hold the insect world in revulsion. Crawlyologists love to look at ants and beetles, while the true scientists study numbers in sterile white rooms. Real scientists leave the study of those tiny, brutal, many-legged sensualists to Walt Disney documentaries narrated by Rex Allen. Frankly, the idea of having my head bitten off after or during a moment of intimacy is not conducive to sexual desire. This is probably why so many crawlyologists are virgins, and the few of them that aren't are either headless or women.

☐ *What are sea monkeys anyway?*

■ The sea monkey is a distant cousin to the lake baboon and ocean gorilla. Long ago, in prehistory, our Neanderthal ancestors discovered fire, the wheel, and the knife—all on the same day. This frightened some Neanderthals. These dissenters decided to go back into the ocean and start over.

Sea monkeys, then, are actually apes that want to be amoebas. You *can* train sea monkeys to do simple tricks, but remember, these tiny apes are very shy creatures, with a paranoid, morbid fear of just about everything. So each trick they learn is just one more step down the road of de-evolution.

□ *I recently saw the movie* Humanoids of the Deep, *which made me wonder, will the dumping of toxic wastes speed up human mutations?*

■ I'm glad you wrote. As you may or may not know, I get most of my information about human evolution from movies, especially those movies that star Doug McClure. My experience has been that movie producers, being relatively unevolved humans, have an angle on human evolution that science just doesn't have. Yes, Hollywood's image of a world gone mad, filled with creatures from the black lagoon, irradiated astronauts, humanoids, androids, and cyborgs, seems pretty much to conform to reality. Even if it doesn't, any movie starring Doug McClure is worth taking a look at.

□ *If Brussels sprouts are sprouts, what do they grow up to be?*

■ Brussels sprouts refuse to grow up. This hideous life form has baffled science for years. Children and dogs, being sensitive to evil, refuse to eat Brussels sprouts. The Latin name for the Brussels sprout is *Vegetus Infernus* or "leafy thing from hell." The first botanist to identify the sprout said it came from the "nether regions." An inept lab assistant changed this to the "Netherlands," and this was further corrupted to "Brussels."

As to their being sprouts, they have been seen sprouting from harmless broccoli, which leads me to believe they might be a type of fungus. Whatever they are, they don't grow up, and neither will you if you eat them.

Chemicals aren't something to mess around with.

□ *Can you describe in scientific terms the workings of a lava lamp?*

■ Yes, I can and I will. The lava lamp contains a small blob of DNA floating in a solution of RNA. These primitive laboratories for life are activated by a simple sixty-watt bulb. Warmed, the blob starts swimming, hunting, yearning—hoping to fill the empty spot inside. Driven by the four horsemen (remorse, terror, greed, and, of course, envy), this blob opts to remain ever unevolved. You see, lava lamps are creatures without the courage to be truly alive. Like college professors, they remain forever in a tepid amniotic limbo. They provide little in the way of illumination and are stared at by adolescents with nothing better to do.

□ *Why do magazines have all those little cards that fall on the floor when you open them?*

■ The publishers of magazines have no concern for the reading public. They want you to think you can get magazines only by subscription or at newsstands. This is simply not true. Those cards are magazine babies, or seeds. Plant them, nurture them, and in six weeks' time you'll have a ripe *TV Guide* or *Reader's Digest*. In two month's time you'll have *Time* or *Newsweek*. After six months you'll have a mature *Playboy*, *Esquire*, or *Vanity Fair*. You must water them carefully, however, or you could end up with a small-town newspaper or one of those real estate guides from the Laundromat. And if you leave them in the ground too long, you'll end up with a novel by Dostoyevsky, in Russian. Happy gardening!

□ *Why is it that when a new box of paper clips is opened some of them are always hooked together?*

■ This is one of the subtle mysteries of procreation and reproduction. Essentially the paper clips thus joined are married. You may have noticed how difficult it is to separate paper clips that are joined together. This is because paper clips are fanatically loyal to each other—and they reproduce like rabbits. Here's a business tip for you people who work in the office stock room: leave the paper clips hooked together for the full length of a clip's gestation period—roughly six and a half months—and I guarantee you'll never have to buy paper clips again. Yes, there are biological mysteries everywhere, even in a desk drawer.

□ *Why do cows all stand facing the same direction while grazing?*

■ Until recently science believed it was because no one, not even a cow, wants to be downwind of a cow. This is true, but there are additional reasons. Grass must be eaten from a specific direction. If a cow goes against the grain, it could suffer severe hoof damage. Other reasons are religious. Cows must all face Wisconsin several times a day. Failure to comply leads to ex-cow-munication and eventually hamburger. Cows do have a language. It only has one word, *moo,* which roughly translates as "cheese food product." Apparently, creating butterfat takes a lot of concentration, and "moo" is a bovine mantra. As conversation, however, it leaves something to be desired—another reason cows choose to stare into space or stick their noses in the dirt, rather than face each other.

□ *Is a dog's mouth cleaner than a human's?*

■ Yes, as a matter of fact, a dog's mouth is one of the most sterile environments in the known universe. You could eat a meal on a dog's tongue if you wanted to. As a matter of fact, many retail companies now offer a highly attractive dog's mouth table service—forks, spoons, plates, even a chafing dish. Now the whole family can enjoy a unique dining experience. Just don't blow in the dog's mouth while you're eating, or your entire dinner will be sneezed across the living room.

□ *How do boneless chickens procreate?*

■ Boneless chickens are incapable of love as we know it. Chickens with bones aren't worth much in the love department either. They are as close to a vegetable as a bird can get. As a matter of fact, the scientific name for chicken is *Aerodynamicus sticklums,* or "fast turnip." The boneless chicken is a freak among freaks. While other chickens peck and scratch, the boneless chicken flops around in the dirt. Science could make a study of the mating habits of boneless chickens, I suppose, but that would mean scientists would have to observe boneless chickens. Maybe watching boneless chickens flop around is your idea of a good time, but science has much better things to do, thank you.

Women often show an early interest in insects.

□ *If fish live in schools and wolves live in packs, what do humans live in?*

■ Nowadays most humans live in apartment complexes. Some of these apartment complexes have their own character, such as the clothing-optional complexes in southern California one hears about. Members of the wild kingdom would be hard-pressed to duplicate the wild times these indoor nudists have in their shag-carpeted pleasure domes. Both humans and animals form their respective communities out of fear, I guess, as is the case with most intimate contact. I've always lived alone: Each of my ex-wives tried but failed to get me to cohabit. A scientist doesn't have time to get involved with schools or packs or marriage.

□ *What's the biggest bird in the world?*

■ That would be Big Bird from "Sesame Street." Before "Sesame Street," Big Bird could only find occasional work as an actor—playing the invisible rabbit in dinner theater productions of *Harvey*, portraying surreal apparitions in the so-called "happenings" of the sixties. Once Big Bird got to play Willy Loman in a highly acclaimed all-bird production of *Death of a Salesman.* But mainly it was hard times. After a dramatic comeback from ten years of hospitalization for acute depression, Big Bird captured the hearts of preschoolers with his convincing portrayal of the whining side of masculinity, mandated by the Corporation for Public Broadcasting in the original funding for "Sesame Street." Big Bird is a survivor, just like Liza Minnelli, only with feathers.

NATIONAL ARCHIVES PHOTOGRAPH

This Scientist has constructed a DNA molecule using only electric train tracks and kerosene.

Chapter Three

Electricity Rears Its Ugly Head

WARNING! The natural forces described herein may well be a form of divine punishment for unatoned sin. The authors encourage the reader to wear thick, rubber-soled shoes while reading.

☐ *Where does electricity go when you turn off the light?*
☐ *Why won't my car start after sitting out all night in Minnesota subzero temperatures?*
☐ *Why is electricity sometimes referred to as "juice"?*
☐ *If sound can't travel in a vacuum, how come vacuum cleaners make so much noise?*
☐ *What do protons and electrons do for a good time?*

Her faith in science allows this woman to trust her luggage to one of the new robot porters.

MORE DAMN ANCIENT GREEKS

As we have shown, Aristotle and his friends spent their evenings shocking each other. Thus, electricity was first used as a plaything. Little did they know what they were playing with. The Greeks also used the lodestone (MAGNET) to stick their dialogues (DYE-UH-LOGZ) on the refrigerator (FRIDJ). It was these twin forces of electricity and magnetism that would join gravity during the Scientific Revolution to end the world as we knew it.

WHAT IS ELECTRICITY?

There are two kinds of electricity. Ordinary electricity powers your vacuum cleaner, radio, and lamp. Extraordinary electricity comes from lightning, magnets, generators, and the tangent galvanometer.

Both these electricities have always been around, but until we discovered sockets and coils we couldn't get at them. Until 1800 electricity consisted mainly of rubbing the cat the wrong way or rubbing your cothurni (SHOOZ) on the carpet and zapping your friends.

Then came Volta's discovery of the electric battery, which he used to make a frog's leg twitch. Frogs' legs had been made to twitch before, but (until Volta) frogs' legs were considered the *source* of electricity. Volta's misguided attempts to stuff frogs into flashlights disproved this theory once and for all. Thus, the way was paved for the first electric motor, powered by cat's fur.

This in turn led to Ohm's Law, which is really too complicated to get into right now. Suffice it to say that Ohm's Law made both AC and DC possible. It led to quantum mechanics, atomic energy, rechargeable batteries, and the liquid crystal display.

And so today we have a whole new world of convenience, at the price of our souls.

WHAT IS ELECTRICITY AGAIN?

Electricity, unlike gravity, magnetism, and atomic forces, does not come to us of its own free will. What electricity there is out there we have to make ourselves. In this way, electricity is much like bread or dirty clothes. It's not here until we make it, and once we have it we have to do something about it.

As to the idea that electricity is our "friend," let me ask you this: Do you live in a home with a gas stove, or do you have an AEK (all electric kitchen)? It's been proven that gas cooks cleaner and is more cost-efficient than electricity. Which keeps you warmer, Grandma's goose-down quilt or that electric blanket? Which lends itself more to seduction, the candle or the light bulb? Sure, electricity can be a force for good. It's given us radio and automatic coffee makers, both of which are necessary to my occupation, but both these things could be easily adapted to coal or nuclear power, and you wouldn't have puddles of dangerous electrons all over the place.

THUMBNAIL SKETCHES OF THE GREAT

Nikola Tesla (1856-1943)—Eccentric Giant from Europe

The Rocky Mountain air crackled with static electricity as the booms from artificial lightning echoed off the foothills. Folks in Colorado Springs clucked their tongues and shook their heads. They resented Nikola Tesla and his experiments. And the arrogant seven-foot-tall figure Tesla cut did nothing to endear him to the natives. They found his aristocratic airs to be just plain unfriendly.

Nor did Tesla find a friend in Thomas Alva Edison. Edison and Tesla were rivals for the electrification of America's cities. Edison championed direct current, while Tesla proposed that cities use alternating current. Alternating current could be raised and lowered in voltage using transformers, but Edison claimed that alternating current was too dangerous for home use. To prove this, he hired neighborhood boys to bring him stray dogs and cats, which he electrocuted in front of reporters. Even these unusual tactics failed to convince the public, and Tesla's alternating current was adopted. Edison went on to invent the phonograph. Tesla took his profits and built a laboratory in Colorado Springs, where he worked on his theory of ethereal energy transmission.

Unfortunately, this latter idea was unsound. Creating artificial lightning was just a dramatic way of creating a nuisance, and Telsa soon bankrupted himself. Returning to New York City, he became increasingly eccentric, eventually communicating only with pigeons. He died, penniless and insane.

If Nikola Tesla were alive today: He would be a character actor in grade-B horror movies.

TO DO AT HOME

SCIENTIFIC METHODS

Experiment number one, being the first experiment, is a crucial one. It will set the tone for the rest of the experiments in this volume. A volume full of experiments is limited in its effectiveness by the preparations we take in making the first of these experiments.

Today's scientist must wage a war against impatience. Impatience is the handmaiden of disaster. Even the simple scrubbing of one's hands must be thorough and convincing. No halfhearted scrubbing can be tolerated if we are to create life in the lab.

Before I undertake even the most rudimentary lab work, I soak my hands in a solution of bleach and laundry detergent. If the work I am to do will be particularly delicate, or if I'm in "that kind of mood," I will soak my hands overnight.

When I was young I would put a bucket of this solution next to my bed, but since this often proved sloppy and unreliable (and since I have eliminated the need for sleep with daily vitamin injections), I devised a system using baggies draped over my hands and taped to my wrists. This is, of course, a painful process. In the morning my fingers are so wrinkled they resemble long prunes. But like so many things we touch upon in this book, this is the price we pay for our devotion to science.

There is a tendency for the perfectionist to spend too

"That is where chemistry comes in."

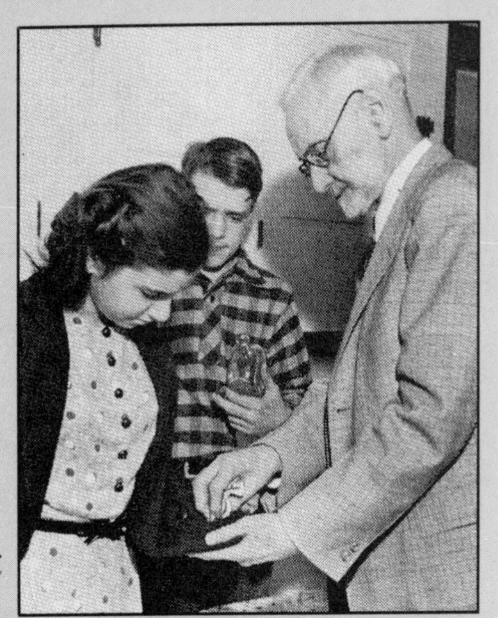

"Now we'll use some alkali to neutralize the acid."

much time in preparation to the exclusion of time spent actually "experimenting." Experimenting is the very stuff of experimentation. We must remember this. It's easy to lose our focus. What with spills to be cleaned up and stopcocks

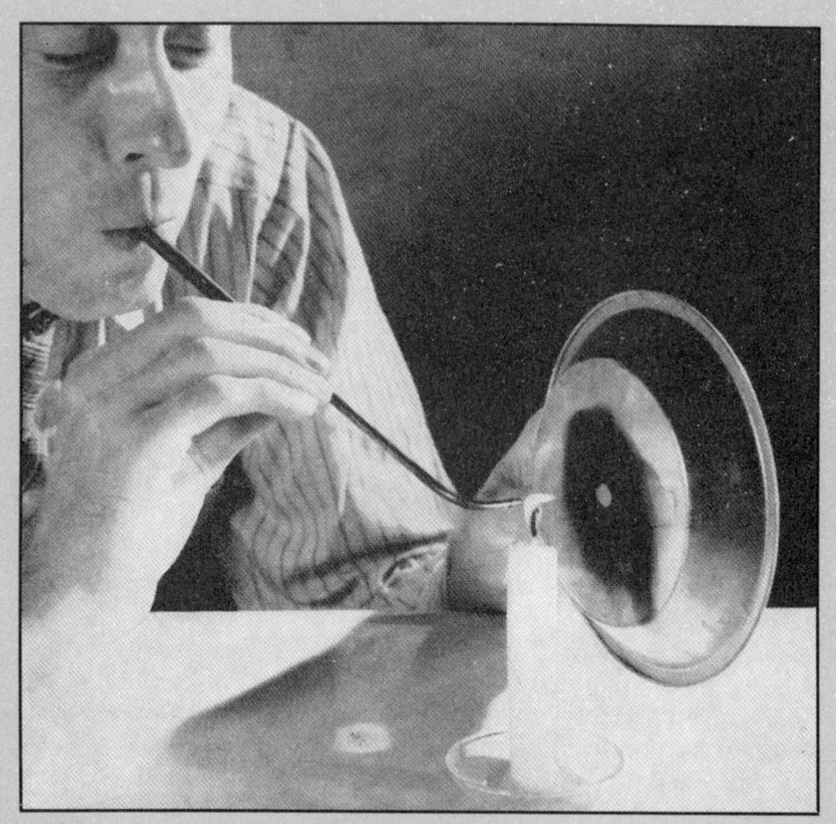

Personal transformation and the transformation of elements happen in exactly the same way.

to be greased—well, you could spend all your time tidying up.

Only through our actions do we convince ourselves we are sincere in our desire to act. We must remember, personal transformation and the transformation of elements happen in the same way. A solution left to boil on the Bunsen burner is exactly like a person sitting in a hot tub. Both share random organic bits floating around in a lukewarm stew. Both are fragile, emotionally and physically. The chemistry that may exist between two people who choose to share a hot tub is similar to the chemistry that occurs between two oppositely charged subatomic particles when they meet. A great deal of energy is discharged and then the glassware, and hot tub, must be cleaned.

THUMBNAIL SKETCHES OF THE GREAT

Thomas Alva Edison (1847–1931)

This hyperactive inventor was born poor and never earned a master's degree in anything. Instead, he worked day and night to invent (among many other things):

- ☐ the light bulb
- ☐ the phonograph
- ☐ the battery
- ☐ the motion picture
- ☐ the loudspeaker paging system
- ☐ the answering machine
- ☐ the doorbell
- ☐ color commentary at sporting events
- ☐ vertical control knobs
- ☐ white courtesy telephones
- ☐ the Checker cab

Thomas Alva Edison had no private life. Draw your own conclusions from that.

If Edison were alive today: He would be a poorly paid engineer for a minor telecommunications company. He would devote his life to the development of annoying new telephone sounds to replace old-fashioned ringing.

□ *Where does electricity go when I turn out the light?*

■ As far as we can tell, it goes to sleep. Electricity, however, can escape unless you keep plugs in every outlet to keep the electrons from leaking out onto your floor. Electrons are highly negative particles, but apart from that we don't know much about them. They're hardworking, I'll give them that much. When they're not making the refrigerator cold, they're turning bread into toast. But how they do this is anybody's guess. Those scientists who say they understand electricity are quacks. I recommend turning your lights out as much as possible. You can't be too careful when it comes to electricity.

☐ *Why won't my car start after sitting out all night in Minnesota subzero temperatures? Do the electrons freeze solid in the battery? If this is true, why don't the electrons in our bodies freeze when the windchill is eight below?*

■ Don't worry. Electrons can't freeze. Electrons can only melt. That's not your problem. As you know, there are two seasons in Minnesota, winter and what science calls "mosquito weather." What causes your car to stop moving are mosquito eggs in the gas line. They're only looking for a warm place to raise a family like everyone else.

There's nothing to be done about mosquito eggs, except to make very small omelets. I would recommend, however, changing the electrons in your car each spring; otherwise you could end up with a sludge of melted electrons in your driveway. Glad to help out.

☐ *Why is electricity sometimes referred to as "juice"?*

■ In the early part of this century, fruit was used in many urban centers as a conductor of electricity. Copper was a better conductor, but copper was in short supply, being used for bullet casings by our armed forces in World War I. So the fruit vendors of the major cities offered their fruit to help electrify our homes and support the war effort. New York City surpassed all others in this undertaking, which is why Manhattan is often called "the Big Apple."

This also helps to explain why our city streets are so often in disrepair. Street crews are removing rotten fruit and replacing it with copper wire. So "juice" is merely a nickname for electricity. The actual residue of electricity is rather dusty in texture.

☐ *If sound can't travel in a vacuum, how come vacuum cleaners make so much noise?*

■ Vacuum cleaners are, in themselves, silent. What makes the noise you find so offensive are the actual particles of dirt and pollution in the space being cleaned. If your living room were clean when you vacuumed, then your vacuum cleaner would make no noise at all. The flaw in all this, of course, is that if your living room were clean you wouldn't be vacuuming. Since there is no such thing as a perfectly clean living room in this less-than-best-of-all-possible-worlds, scientists had to prove this hypothesis by vacuuming in outer space, itself a perfect vacuum. Space is also incredibly clean. Astronauts reported that even the most powerful, poorly maintained vacuum cleaners made absolutely no noise in space. Millions of your tax dollars went toward proving this.

□ *What do protons and electrons do for a good time? Where do they do it?*

■ Electrons go to any popular subatomic particle bar to meet protons. Being negative by nature, they consume large amounts of alcohol in a vain attempt to charge themselves up. What happens after that is none of my business. Protons, being many times more massive than electrons, frequent trendy spas and gyms. These places are a good place to show off what they want other charged particles to see. Uncharged particles, or neutrinos, also have their own gathering places, although they are often the victims of harassment by immature particles who are insecure about their own valence.

SUMMARY OF CHAPTERS ONE, TWO, AND THREE:

WHAT WE'VE LEARNED SO FAR

The ancient Greeks had many wrong ideas. Galileo proved that gravity originated from the base of the Eiffel Tower. Sir Francis Bacon invented the diary. Issac Newton isolated gravity. Evolution tells us that everything is made of something else. Electricity is responsible for all the evil in the world today.

REVIEW QUIZ

1. Many wrong ideas were held by the ________ Greeks.
2. In your own words, describe a typical ancient Greek.
3. Now describe a typical ancient Greek in *my* own words.
4. Was Aristotle ever the Father of Anything? If so, what?
5. Complete the following syllogism.
   The ancient Greeks had many wrong ideas.
   Aristotle was an ancient Greek.
   Therefore: ________

NATIONAL ARCHIVES PHOTOGRAPH

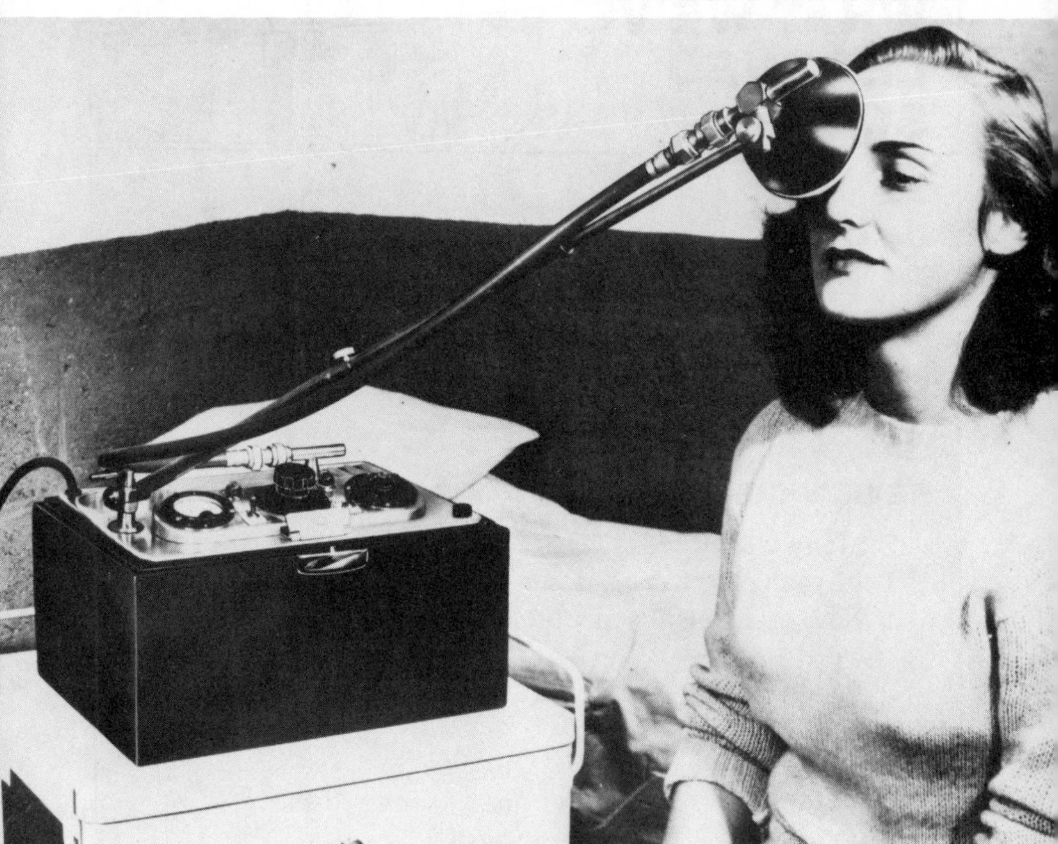

This foolish woman seeks "thrills" by transmitting electric "waves" directly into her brain.

Chapter Four

How Light Got Turned On

A PROMISE: The truth shines like a beacon, dispelling the shadows of ignorance. If you can't stand bright lights, wear protective eyewear.

☐ *Why do fluorescent lights hum?*
☐ *How do you tell a quark from a lepton?*
☐ *What would happen if the speed of light were only sixty miles per hour?*
☐ *Why do objects become shorter and wider as they approach the speed of light?*
☐ *Is a* quantum leap *some kind of track event, like the broad jump or pole vault?*
☐ *Why is the speed of light only 186,000 miles per second? Can't science do better than this?*
☐ *What does* polarization *mean? Does the North Pole have anything to do with it?*
☐ *I want to build a time machine and travel back to a better year. Any suggestions?*

Neanderthals huddled around the fire.

WHY LIGHT?

Most scientific breakthroughs, like musical fads, travel in bunches, or waves. After the Industrial Revolution and the Age of Newton, science was nearly drowned by the wave of new information. It lay like a sick fish on the shore of progress, flopping and gasping for air. But the fish of science wasn't as sick as it looked. It had a headache and chills, sure, and an achy feeling, but it was growing as it flopped. It was assimilating. And it was about to create a *paradigm* (PARE-UH-DIME), a Greek word meaning, "Oh, another revolution." It was about to discover the speed of light.

PREHISTORIC LIGHT

In many ways, a primitive tribe of Neanderthals (KAYV-MEN) resembled a Cub Scout troop. They huddled in terror around a campfire and whiled away the lonely nights with ghost stories. Of course they didn't have merit badges or uniforms, and if there was a Cub Scout manual around they probably used it for fuel, but they could tie a granny knot with the best of them and knew how to start a fire with nothing but a little stick and a handful of tinder (MATCH-EZ).

Sooner or later, one of these primitives got tired of huddling. Perhaps he heard the tribal chief tell those stale old creation myths one too many times. We don't know. We only know his attention wandered. He got up and walked around.

Ironically, it was when this primitive stopped paying attention that he first noticed something. He noticed lights in the distance, the lights of other tribes huddled around their fires. And some of those fires were bigger than his. At that point two questions occurred to this bored, bright primitive:

1. How fast does it take that distant light to reach me?
2. If I steal that fire, will I be warmer?

So he got his tribe together and stole that bigger fire. Thus was born modern civilization. And thus was born a new topic of conversation—something besides religion, weather, and the price of mastodons. That topic was the speed of light.

BIRTH OF MODERN LIGHT

The story you have just read is mere conjecture, but it is probably close to the truth. At least nobody can prove it isn't. We can see in that epoch-shaking moment modern man (HOMO-SAY-PEE-EN) emerge in all his glory—the greed, the envy, the politics, the curiosity about the speed of light, even the brief attention span.

There's no real way to tell how long light has been around. The old song tells us to "hitch our wagon to a star," but many

THUMBNAIL SKETCHES OF THE GREAT

Antoine Lavoisier

The ancients believed that if something burned, it was on fire. We now know it's not that simple. In the 17th century, the Phlogiston (FLO-JISS-TUN) Theory was developed. Briefly stated, this theory held that if something burned, it was on phlogiston. This was a more complicated way of saying what the ancients said. Until the four-element theory (earth, water, fire, and air) and phlogiston (fire) could be thrown out with the bathwater, modern science just had to sit in the waiting room.

It was Lavoisier, the French inventor of mouthwash, who turned fire around. He removed oxygen from air in a classic experiment; he found he could not only suffocate small animals this way but make fire go out as well. A compulsive arsonist, Lavoisier took time out to draw up the first "table of elements": Hydrogen, Dynel, Carbon, Oxygen, Phosphorus, Neon, Milk, Lime, Soda, Scotch, Gin, Dirt, Iron, Zinc, Copper, Lead, Silver, Fire, 2.2.4 Trimethylpentane, and Water. He had no proof for his ideas and his elements were mainly wrong. He was put to death on the guillotine, but not before earning the title, Father of Modern Chemistry.
If Lavoisier were alive today: He would be in prison.

of those stars are old and burned out, and we'd be hitching our wagon to nothing. So what is light? How do we measure it?

Until the Industrial Revolution, nobody had any answers. The reason? Until the Industrial Revolution, it was too dark to see. Aristotle bears much of the blame for this. When he put the earth at the center of the universe, light just couldn't function the way it should. And light wasn't trusted back then the way it is today. The Dark Ages were called the Dark Ages because people stayed indoors. And who can blame them? Outside were plagues, eclipses, wars, the Inquisition, and more popes than you could shake a stick at.

EARLY MEASUREMENT OF LIGHT

It was that old fraud Galileo who first tried to measure the speed of light. He bought a lantern and hired an assistant (a street urchin named Guido, who, by accepting this position, became the first person in recorded history to have a job). He placed his assistant on a hill three miles away. Galileo then shone the lantern at this distant hill. When the assistant saw the light, he yelled back, "I see it." Galileo then measured the time elapsed between shining the lantern and his assistant's shout, giving him the first crude measurement of the speed of light. This worked out to be roughly fifty-five miles per hour. Though this earned Galileo the title "Father of the Speed Limit," he was forced to fire his assistant (who thus became the first person in recorded history to receive unemployment compensation), and the true speed of light remained a question mark.

Is Light a Wave or What?

It has been known since the invention of The Shout in 200 B.C. that sound waves are created by vibrating the air. This is still true today, as you can prove for yourself by chopping down a tree in a forest when no one is around to hear it. Until the Industrial Revolution, the forests were the noisiest places in the universe, home of many famous battles, large animals, ogres, and Robin Hood. Then Lavoisier proved that sound can be very quiet in a vacuum, but light had no trouble passing through a vacuum at all! This led scientists to believe that light, unlike sound, might not be a wave after all.

Early scientists believed that light, like disease and thought, traveled through the ether. If true, they theorized, you should feel the breeze of the ether when light goes by. Many scientists were severely sunburned trying to feel this wind, before they wised up and began to use small animals. These animals, usually rabbits, were set out in groups to catch the first morning light. When they weren't knocked over by the ether, the theory was finally discarded. The rabbits, burned pink by the experience, were given away to children. This so-called Ether Bunny tradition continues to this day.

WHAT?

Once the ether theory was discarded, science was left in the dark. Fed up with its own metaphysics, it got tired of the question "What is light?" and turned to the easier question "How fast is light?" After all, it was reasoned, one doesn't need to know how a car is put together to obey the speed limit.

The speed of light was finally measured with the invention of the motion picture (MOO-VEE) in 1897. The first movie was made by Thomas Edison and the French magician Georges Méliès. It starred a very young Charlie Chaplin. In the audience at that first screening was Albert Einstein, who also produced the movie.

To find the speed of light, he simply counted the number of frames in the film and the number of people in the audience. The first time the audience laughed at Charlie's antics, he noted the time elapsed in the running time. He then subtracted this number from the total running time of the film and multiplied that figure by the number of kernels of popcorn that had been consumed. This gave him a speed of 184,000 miles per second, which worked out to be surprisingly accurate.

Since that time, movies and light have become more complicated. We know now that light travels in bunches of wave/particles called *photons*. To understand the photon, think like this: if space is the ocean, then light is a kind of surfer/wave combination. Of course, today we don't think of surfing as work, but we must remember that light wants a good time like every other force of nature. And you can't have fun in the darkness, as Einstein has proven. Fun in the darkness is merely research. Ask any of my ex-wives.

Dr. Science's ex-wife, before.

THE THEORY OF RELATIVITY

Modern equipment has shown us that light travels at 186,000 miles per second and that light is part of the family of electromagnetic waves. Only light waves are visible, however; the rest travel in the dark. Since the invention of electricity, most light has been artificial. We find true light these days only as a secret ingredient in diet soft drinks.

It was Albert Einstein, high school dropout and Hollywood mogul, who showed us these truths, and many more, with the Theory of Relativity. This proved, scientifically, that the world is complicated and confusing. Many people were disturbed by his findings, but as Einstein himself said, "If you can't stand the heat, stay out of the linear accelerator."

And what is this theory?

1. *You can't get there from here.* This elegant principle, a folk truth for centuries, freed people from the agony

Dr. Science's ex-wife, after.

THUMBNAIL SKETCHES OF THE GREAT

Max Planck—A Quantum Leap Ahead of the Competition

Max Planck knew he was in trouble. His Black Box was starting to smoke. The Black Box was the one he had been counting on to manufacture "quanta," a new confection he hoped would become the rage at Vienna's after-opera sidewalk cafes.

After quickly switching off the power to the box, Max opened the door of the box, and a bright beam of light shot out. This beam, a steady stream of photons, was actually the first laser. Of course, Max Planck had no idea he was looking at the first laser beam. He muttered curses under his breath as he lowered the box into a bucket of water. Then something startling happened. As the overheated box hit the cool water, instead of hissing and bubbling it began to make a dreamy kind of music, and spiral patterns of light swirled across the ceiling. Thus the first laser light show was born.

Fortunately, Max Planck was able to understand what had happened and to analyze the photons his Black Box produced. These he referred to as "quanta," a sentimental way of recalling the happy accident that made Max Planck the father of what we now know as "the Quantum Theory."

If Max Planck were alive today: He would own a head shop in the Haight Ashbury District of San Francisco.

of leaving their homes. Thus today we enjoy hours of television programming; with modern telecommunications we can even shop at home and never go outside at all.

2. *If you go fast enough, you will disappear.* Once you go past the speed of light, you become darker and darker, until not even your closest friends can recognize you any more.
3. *If you sneeze, fart, burp, hiccup, and cough at the same time you will die.* Though this has never happened, it is theoretically true. Einstein's mathematical proof of this maxim runs twelve pages long, and is really too complicated in scope for this present volume.

THE UNIFIED FIELD THEORY

The theory of relativity put everything in its proper perspective, and it only remained to bring out the theory in a simplified package for the layman. In a process called

novelization, the Theory of Relativity was reduced to the Unified Field Theory. While easier to grasp than relativity, it was still very complicated and scientific. It can be paraphrased roughly as "Everything's beautiful in its own way."

As we shall see, this is the precept of the New Science. When everything is beautiful in its own way, we can discard guilt from our lives and truly live a "lifestyle" that the world can envy. We can dispense with ideas and accumulate data. We can do anything we want to, and it's OK!

We must be careful, however, from our clean vantage point in the future, not to impose modern values on ages past. When we say today that light "words," it doesn't mean that dark doesn't work or that darkness is lazy light. It only means that people who sit in the dark are lazy. This accounts for the popularity of the day shift in the modern workplace.

For most of us, perhaps, the speed of light is a useless fact. But if, in olden times, 186,000 miles per second seemed fantastically high, today it seems disappointingly low. We can splice a gene and blow up the world, but light stays constant at a measly 186,000 miles per second.

☐ *Why do fluorescent lights hum?*

■ Because they don't know the words. Unlike incandescent lights, fluorescent lights can remember only the melody. Most are born with perfect pitch, and harmony is second nature to the long white tubes, especially harmonics of sixty cycles per second. Incandescent lights, first invented by Edison, are tone deaf but are eloquent speakers. It's not uncommon for a 40-watt bulb to recite the Gettysburg Address verbatim. And the new extended-life bulbs can rattle off the entire Constitution. Like anyone in show business, both fluorescent and incandescent lights are subject to burnout. Once this happens they join the scrap heap of burnouts and move to northern California.

□ *How do you tell a quark from a lepton?*

■ Easy. A quark is a subatomic particle; lepton is a type of tea. Tea can keep you up nights, and so can quarks, if you're a physicist, but there the resemblance ends. Quarks have only a theoretical mass, and tea comes loose or in bags. If quarks came in bags, it would make a physicist's life much easier. But quarks don't. In the case of the quark that we call the boson, the Heisenberg principle proves that even observing the boson destroys the boson, which is not to be confused with bison. You can look at bison all day without destroying anything. Bison are large mangy buffalos. They live in Wyoming and don't drink tea at all.

□ *What would happen if the speed of light were only sixty miles per hour?*

■ As we approach the speed of light, the aging process slows down. So, if the speed of light were sixty miles per hour, we would have even more people speeding, especially older people trying to stay young. As a matter of fact, physics would demand that we go faster than the speed of light. The safest thing is to drive at a steady sixty to keep time and the highway patrol off our necks. Airplanes would become obsolete in this slow light world, because you would be going so fast, relatively speaking, that you'd be back before you even left. This would make business trips unnecessary and lead to economic collapse. So, to answer your question, life, if the speed of light were sixty miles per hour, would be youthful, fast, and dark.

□ *Why do objects becomes shorter and wider as they approach the speed of light?*

■ There are two different kinds of light here, the light that fills our days and the light that fills our beers and diet sodas. The objects that become shorter and wider are those that consume too much light beer. The so-called "couch potato syndrome" could be more a side effect of gravity than of light, though the light emitted from a TV set seems to have an adverse effect on weight. TV light, or, as science calls it, "stupid light," seems to create an urge in couch potatoes to drink gallons of light beer. Why, we don't know. Stupid light contrasts with smart light, which is the intelligent radiation we get from the sun and Everready batteries. When we approach the speed of smart light we don't get shorter and wider; we get dark, bump into things, and fall down. So, if you plan on breaking the light barrier, I advise you not to. Turn on the TV and crack a couple cold ones. You'll be fat, but you'll be safe.

□ *Is a* quantum leap *some kind of track event, like the broad jump or pole vault?*

■ The quantum theory says that energy exists in units called *quanta*. The quantum of light is the photon; a quantum of nuclear energy is the meson. A half quantum is called a *pint*, and half of that is a *microcup*. From there it's broken down into teaspoons and tablespoons—all theoretical quantities, of course. Four quanta do make a gallon, but unless you're painting the living room, a gallon of light is more than anybody needs. When scientists say they've made a quantum leap, it's merely jargon that means they went outside to catch some sun. Scientists are very pale as a rule, and when we go outside we tend to make a big deal out of it. Hope this answers your question.

□ *Why is the speed of light only 186,000 miles per second? Can't science do better than this?*

■ Yes, you're right. It's a disgrace that light goes only a measly 186,000 miles per second, but physicists are working on the problem. There is already a prototype vehicle that goes 200,000 miles per second, but the headlights shine at only 186,000 miles per second. This is equivalent to driving down the freeway the wrong way with the headlights not only *out* but actually chasing you down the road. This is why so many scientists today no longer own a driver's license.

□ *I'm trying to decide between two kinds of sunglasses: polarized and nonpolarized. What does* polarization *mean? Does the North Pole have anything to do with it?*

■ Polarization of light is similar to the pasteurization of milk or the simple boiling of water to weed out bacteria. Polarization takes out bad light that hurts your eyes, leaving the good light so you can enjoy your summer fun. How can you tell which sunglasses are polarized? Sunglasses shaped like ducks or with tiger stripes and polka dots usually are not polarized. Polarized sunglasses are usually made by Italian designers and cost as much as a three-piece suit. Personally, I don't own any kind of sunglasses. I just stay inside all the time. It's so much easier, and there's no risk of sunburn.

□ *I want to build a time machine and travel back to a better year. Any suggestions?*

■ Yes. According to my calculations, a time machine will be feasible by the year 1999. My suggestion is to wait until then, buy one of the cheaper models at Radio Shack, send yourself back to the present time, and give yourself the time machine. Now that your present self has the time machine, go forward in time and give the time machine back to yourself before your future self goes to Radio Shack. This way you won't have to pay for it. As to which year is best, that's a subject open for discussion among you and your past and future selves. I can't recommend one year over another. I'm a scientist, not a travel agent, and I don't have time for vacations.

NATIONAL ARCHIVES PHOTOGRAPH

A prototype of an early videogame is tested in Bethesda, Maryland, 1948.

Chapter Five

How Work Works

Unemployment is often disguised as self-employment. Learn the telltale signs of the chronically unemployed, as well as guilt-free ways to become unemployed yourself.

☐ *Are eggs the only food group that work?*
☐ *How can we tell if sour cream has gone bad?*
☐ *My home computer sits idly in the corner. How can I modify it so it will brew a decent cup of coffee?*
☐ *Why is there never a working ballpoint pen attached to those chains you see hanging around banks?*
☐ *What would happen if you cleaned your self-cleaning oven?*
☐ *Why does freeway traffic slow to a crawl whenever you see a car abandoned on the shoulder?*
☐ *My parents told me that dollars don't grow on trees. Could you tell me what sort of life form produces dollars?*

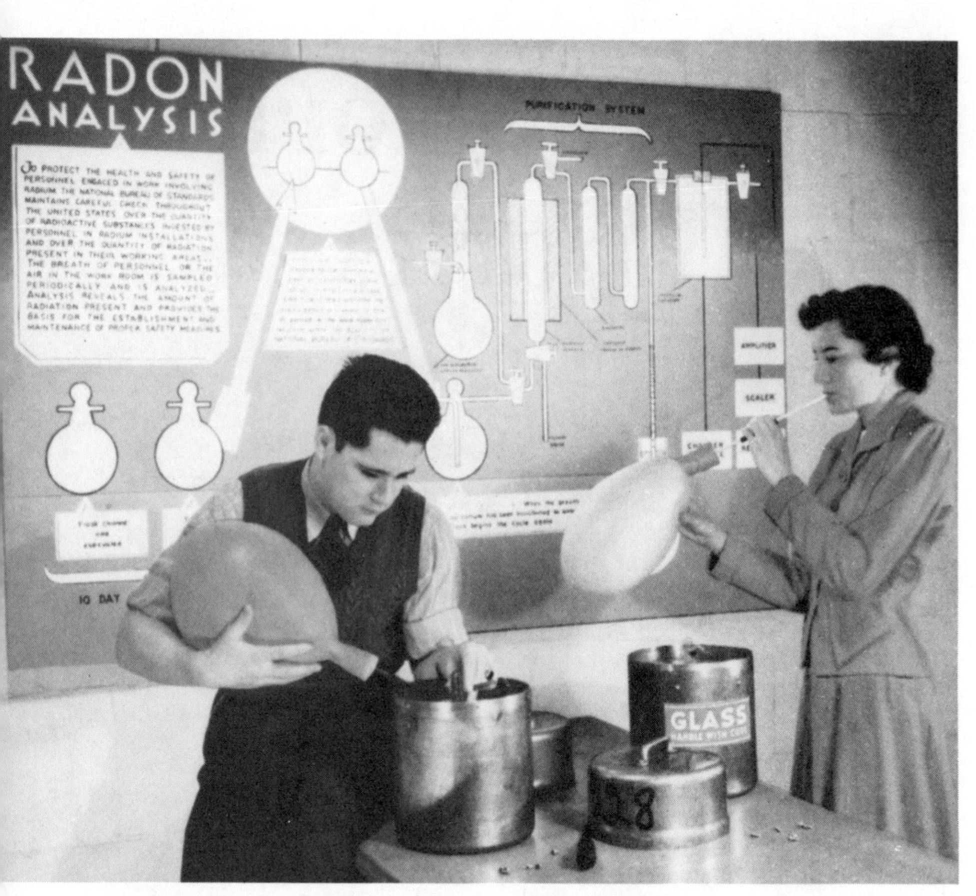

Students inhale radioactive gas for kicks. Not only will their grades suffer, but they may not live to see the year 2000.

WHAT DO WE MEAN BY WORK?

We are not discussing work in the so-called scientific sense. Work here is not the product of a force and a distance moved by the point of application in the direction of the force, nope. We're talking about ordinary Joes and Janes shagging a bus to a nine-to-five. You probably hate your job. That's normal. But you might not know that by having a job, you are obeying certain Universal laws.

UNIVERSAL LAWS

For centuries the world had been divided between peasants and royalty, but there wasn't much difference between them. They both had fleas, and they both lived in what is known as the Holy Roman Empire. Unless there was a Crusade, they seldom left the castle.

Then in Newton's time, the Universal Law was discovered: Time Is Money. Once time equaled money, life ceased to be a hobby and became a job. There was a positive side effect, however. Once we had work, its opposite, "fun," began to have shape and definition. Fun was Not-Working. This gave us the corollary to the Universal Law: If You Are Making Money You Are Working; If You Are Losing Money You Are Having Fun.

LIGHT AND WORK

Though the true speed of light was not determined until the 20th century, light in its present form was tamed with the discovery of the socket in 1775. Electric lights began to appear everywhere, and we began to see how ugly things really were. This led to an orgy of cleaning and dusting, which led in turn to exhaustion and the condition known today as self-doubt (INTRO-SPEK-SHUN).

This led to long walks at twilight, heavy drinking, and brooding. Our days and nights had become brighter, and we had to find ways to fill them. We became bored with our old surroundings. We replaced the horse with the steam engine. We built taverns, factories, and railroads. Suddenly there were jobs everywhere: electricians, plumbers, trash collectors, temporary agencies, even guidance counselors (named for Guido, Galileo's hapless assistant). And thus modern work was born.

SCIENCE TAKES A WRONG STEP

And where was science when work first reared its ugly head? Science had taken to its bed with a headache, and the name of that headache was *electricity*. Frightened by the light it had tamed, and too exhausted to measure it, science, like Marcel Proust, turned inward.

Once work became a tangible thing in the world, it became obvious there just wasn't enough to go around. This led to "self-employment" (LOAF-ING), a Greek word meaning "watching others work." Many scientists became self-employed, thinking thereby they could fool others into thinking that what they did was work. Thus the Victorian

THUMBNAIL SKETCHES OF THE GREAT

James Watt—Inventor of the Steam Engine

The winter of 1810 was especially severe in Scotland. James Watt tried to warm his hands from the steaming teakettle and thought, "There must be some way to make steam do useful work." Lost in thought, he failed to notice that the steam had badly scalded his hands.

Later that afternoon, in the physician's waiting room, he saw a bent old man turning a heavy piece of pill-grinding machinery. When he asked the doctor what the old man was doing, the doctor told him to mind his own business. This lit a spark of resentment in the mind of James Watt.

It was months later, after his hands had healed, that Watt built the world's first steam engine. It was a flimsy contraption and took up most of the backyard. On the day he decided to test his invention, he lit a fire under it and began the slow process of building up pressure. When it finally had a full head of steam, it exploded, leveling the Edinburgh row house Watt shared with other desperate families.

If James Watt were alive today: He would be a minor executive in a large corporation, drinking heavily, bitter that those with less seniority had been promoted and he had not.

TO DO AT HOME

THE ELECTROMAGNETIC FIELD THEORY

Nothing beats an electromagnetic field for creating that certain "ambience" everyone seems to want nowadays. Any fashionable restaurant depends on large electromagnets that bathe the dining area in a "flux" that enhances the DEC (Dining Enjoyment Coefficient) of its patrons.

In restaurants these electromagnets are often disguised as dishwashers and walk-in coolers. When a restaurant goes out of business, many of these dishwasher- or cooler-sized electromagnets can be picked up at bargain prices. Pick one up. Take it home.

From the moment you switch it on, you will see all the metal objects in your living room line up, as if by magic, facing the electromagnet! And you will feel the "vibes" of the electromagnetic field. These vibes often reveal themselves as either a general feeling of well-being or acute paranoia. An insensitive subject will equate them with the normal feeling that comes with everyday life, but the truly perceptive student of electromagnetic theory will know vibes when he feels them.

And a true scientist has a way of knowing when an experiment is causing more pain than it's worth. When the pain/gain ratio gets too large, a scientist has a sixth sense that triggers a silent alarm. What this alarm is we don't know.

Even batteries can prove deadly, if misused.

We know that it's not a high-pitched whine, because laboratory animals do not seem to react when a scientist "freaks out."

Certain experiments indicate that there is a hormone produced by the *pons illuminati*, a walnut-sized gland just above the inner ear. When this hormone induces a state of uneasiness, the scientist shuts down the experiment. In healthy scientists, this is a marvelous survival mechanism.

In unhealthy scientists, the survival mechanism malfunctions. The pain/gain ratio grows top-heavy. Then we all suffer. History abounds with examples of experiments that would have been better off being shut down right after they began. The polyester shirt. The Nehru jacket. Cheese food by-products. Naugahyde.

So the main function of an electromagnetic field in the home is to see if our sixth sense is functioning. If it isn't, our neighbors will soon complain, because all the metal objects in their home will begin to face our direction. If they don't complain, or if we ignore them, we will probably go insane. It's that simple.

So the themes of this experiment are common sense and sensitivity to feedback.

Age gave us the softer, wistful sciences: economics, sociology, phrenology, psychology, biology, medicine, and subatomic physics.

Real science took a backseat for most of the 19th century, resting on its electric laurels. This trend was to have a profound effect on modern times. These so-called "scientists" are still around today. If you meet someone who is not working, chances are he or she is self-employed as a social scientist, therapist, or lawyer. None of these professions were necessary until the Industrial Revolution. Today all of us need someone to tell us what to do.

Karl Marx

It is ironic that Karl Marx was the first human being to "shed light" on work, ironic because his own personality was dark, and he never worked a day in his life. By the Universal Law then, was he having fun? No. He was a deep thinker and didn't know how.

He lived several doors down from Charles Darwin. After seeing the fame surrounding that famed naturalist, he decided to try his own hand at an evolutionary theory.

By nature, Karl was a finicky man. Small squirmy things made him ill. So he avoided bugs, birds, and worms and made man the object of his study.

First, he invented the candle. Then he put the kids to bed and worked through the night. By dawn he had invented dialectical materialism (GAWD-LESS-NESS), which says that everything is real but some things are more real than others. Bosses, in other words, are more real than workers and will be until workers control the means of production.

This idea was ignored for a century by baffled workers, until it was forced down their throats by psychopathic Russian students in the 20th century. This led in turn to a Cold War (MEW-TEW-AL DE-STRUC-SHUN) with America, a nation founded on capitalism (GREED), invented by

Henry Ford in 1910. Capitalism, too, says that some things are more real than others. In capitalism's case, products are more real than people.

As to which is more real, capitalism or Marxism, we don't know. We only know that work itself is as real as light, but we have not yet discovered the speed of work. Science is working slowly on that problem.

☐ *To measure temperature, many people fry eggs on the sidewalk. Is this the only food that works?*

■ Eggs are the food of choice for the scientific measurement of heat. Bacon and hash browns, even hamburger, can be used, but they leave stains on the concrete that are difficult to remove. Eggs are also used to indicate humidity and barometric pressure. If the egg is wet, for example, science can infer that it is raining. If the egg explodes, barometric pressure is high; if it implodes, pressure is low; and so forth. The main difficulty with eggs, of course, is their tendency to stick to the cement. To avoid this problem, most meterologists carry spatulas with them on hot days and have had their driveways coated with Teflon.

☐ *How can we tell if sour cream has gone bad?*

■ This is a complicated question. Sour cream is, of course, an early by-product of artificial intelligence research. The original trade name for sour cream was Mushy Plastic Brains, a not very appetizing name, although calling a product "sour cream" doesn't stir the taste buds much either, frankly. It's hard to tell when sour cream first goes bad, but there are warning signs—when the sour cream starts hanging out on the corners with cottage cheese and Brie, when sour cream comes home at 4:00 A.M. smelling like a brewery, when sour cream is dull, listless, and sarcastic. Any or all of these symptoms could mean bad sour cream, and you should seek family counseling at once.

☐ *My home computer sits idly in the corner. How can I modify it so it will brew a decent cup of coffee?*

■ Chances are you're not looking at a home computer sitting idle in the corner, but a coffee maker. Today's modern coffee makers do bear a strong resemblance to computers, and you're probably not the first confused user to mistake one for the other. A good experiment to establish the identity of the appliance in the corner involves filling the basket just under the top with coffee grounds, then pouring cold water into the grill on top. If the fuses blow and you end up with a floppy disc covered with coffee grounds, you're on the wrong track. But if, moments later, you find yourself enjoying a fresh brewed cup of coffee, well then, my hunch was correct. Good luck!

☐ *Why is there never a working ballpoint pen attached to those chains you see hanging around banks?*

■ Once imprisoned, the ballpoint pen soon loses the will to write. Even if released, it develops a deep-seated pattern of skipping and blotching that will soon send it back to the chain gang. Repeat offender ballpoint pens are sent to the post office, where they become federal pens. If not watched carefully, these pens will hang themselves from the writing desks. To prevent this, postal workers are particularly attentive, which is why they so often seem to be standing around staring at the pens. To avoid contributing to this destruction and unwholesome situation, I use a high-resolution dot-matrix printer when I write, and I recommend you do the same.

□ *What would happen if you cleaned your self-cleaning oven?*

■ You would be playing with forces beyond mortal comprehension. To bypass the scientific impossibility of an oven that self-cleans, stove makers made a pact with Satan. Each self-cleaning oven contains its own demon, which resembles a heat-resistant jackal. This demon, which the Egyptians called Ah Mah Nah Raynj, or "ugly dog on fire," dwells in your oven and eats your leftovers as you sleep, right down to the broiler grease. If you disturb this creature as it feeds, you are not only placing your own life in danger; you could be unleashing the very Beast mentioned in the Book of Revelations. If I were you, I'd throw away your oven and eat only at restaurants. If you must cook at home, use a hot plate.

□ *Why does freeway traffic slow to a crawl whenever you see a car abandoned on the shoulder?*

■ This phenomenon, called *gawking* or *rubbernecking*, is actually the result of posthypnotic suggestion by your automobile. A kind of languor, a wistful quality (what the French call *tristesse*) overcomes your car when it sees an empty car on the freeway. It's the same feeling a man gets when he sees sunlight on a woman's hair or a woman gets when she sees Richard Gere play the trumpet. The nurturing instincts emerge, and your car will stop at nothing to get you to stop. The radio will play songs with lines like "slow down, you move too fast," and "stop in the name of love." If you can withstand the mental commands of your vehicle, you are stronger than most people. The combination of internal combustion and love is very hard to resist.

□ *My parents told me that dollars don't grow on trees. Could you tell me what sort of life form produces dollars?*

■ Dollars are produced by a form of bacteria that reproduces only in a dark, damp, tomblike environment. This is why our national mints are such thick-walled, sarcophagal affairs. It's the U.S. Treasury's job to make sure that the bacteria don't reproduce too quickly, because that would cause inflation. If they thrive poorly, that causes a depression. If they mutate, we end up with counterfeit currency that must be eliminated before it infects the whole money supply. So you see, managing the nation's currency is mainly a job for microbiologists, not botanists. Your parents were right.

SUMMARY AND CONCLUSION: PART ONE

WHAT WE'VE LEARNED SO FAR

Work and light are pretty much the same thing. Subatomic particles are too teeny. Einstein knew what he was talking about. Biology and psychology are not real sciences. We're better off today than we ever were.

REVIEW QUIZ:

1. (True or False) Primitive Neanderthals discovered light beer.
2. Time is ________ .
3. Is Karl Marx a good example of human evolution? Why or why not?
4. Is light a wave, or what?
5. Does light "work," and if so, how much does it earn per hour?

CONCLUSION

It must be apparent by now to even the most foolish of our readers that the discovery of electricity and light was a real Pandora's box. Once we sat in the dark like slugs, and were content. But the invention of artificial light brought insomnia with it. We stayed up late thinking and brooding. We asked ourselves, "Who am I?" and answered ourselves, "A bunch of teeny squirmy cells and atoms all stuck together." So people today are disgusted by the fact that atoms and cells are exploding and oozing inside them as they walk around. That's why we get jobs, to take our minds off our mental disgust.

Work has come a long way since the Industrial Revolution. Henry Ford's assembly line made work more efficient, freeing factory workers from the drudge of enjoyable labor. Ray Kroc brought the assembly line process to food, thus freeing teenagers in uniform from the drudge of cooking real food. After World War II, William Levitt brought this same process to architecture, creating suburbia—making houses as fast as people could move into them, thus freeing people from the drudge of living in a home.

As we will see in our next chapter, this is all part of the New Science: Science Now. New Science has made fun possible again. New Science means that none of us work if we can get away with it. We have created "interfaces" (IN-TER-FAY-CEZ) between our public and private lives. *People Magazine* brings the private lives of celebrities right into our living rooms. Commercials have turned our half-felt private mundane wants into a slick public orgy of gluttony and ignorance! If, today, we still have cottage industries who hold that craft and design are more important than efficiency, they are mere holdovers from the Dark Ages.

NATIONAL ARCHIVES PHOTOGRAPH

Even a child's game like "Pin the Tail on the Donkey" becomes work when Science steps in.

Part Two

Science Now

The New Science

NATIONAL ARCHIVES PHOTOGRAPH

INTRODUCTION TO PART TWO

Old Science had much to give us. It gave us a cornucopia of amazing facts: pigeons are the offspring of the union between escaped parakeets and rats, for example. Old Science taught us that Channel One is where television personalities go to die. It told us that all vegetables are deadly poisons, that scientists cannot be hypnotized, and that chickens do indeed have lips. Where would a data bank be without this kind of data? A Trivial Pursuit game would be a dull affair without Old Science. Yes, Old Science had been a true friend, but even the most trusted friend must be put to sleep.

Once we knew what the universe was, how light and electricity worked, once we understood the process of evolution, what then? Did your new understanding help you get a job? Or inner peace? Or even a parking space? Has the Old Science helped you find the conditioner that's right for you? Of course not.

The Old Science has died, and from its ashes arises the New. There in the embers, like the Phoenix, or popcorn, the

The following table displays the differences between Old and New Science. On which side of this table would this table itself fall?

Old Science	New Science
Subatomic Particles	Waxy Buildup
The Sound Barrier	Compact Disc Players
Accumulation of Facts	Opinion Polls
Monographs	Letters to the Editor
Pasteurization	Taste Tests
Telephone	Walkie Talkie
Krypton	Kryptonite
Data Base	Photocopy
Geology	Real Estate
Anatomy	Aerobics
Simultaneity	Synchronicity
Albert Einstein	Wilhelm Reich
Aristotle	Zorba
Sir Isaac Newton	Dr. Who
Lavoisier	Dr. Phibes
Charles Darwin	Dr. No
Origin of Universe	Creative Accounting
Life on Earth	Lifestyles on Earth

New Science rears its handsome head, flaming and popping, ascending to the heavens!

All right, let's calm down. What are the differences, you wonder, between New Science and Old? For one thing, Old Science merely described nature, by telling us what chemicals were in our food. New Science goes one step further and *puts* chemicals in our food. Any dope can put lemon juice in a lemon, but to put lemony taste in a soft drink takes brains. And to find a lemony taste that is not only artificial and inexpensive but non-lethal as well, takes more than brains. It takes determination. And guts. And money. Lots of it.

Grant money is another important difference. The unselfish scientist who works at home is a dying breed. When Big

Money entered the scene, New Science entered as well. Science had to go the way of banking, the military, or cosmetology. When the tough needed science, the New Science saw dollars.

And this is the most important difference. Old Science was a penniless wimp. It would agonize for years over methodology and truth. New Science has no interest in truth, only in facts. Old Science would hang out with poets and painters; New Science is not friendly. It can't afford to be. The New Scientist walks alone, beset by anxieties, but putting a brave front to a colleague or a grants committee.

But we're getting ahead of ourselves. Before we can roll up our sleeves and get to work, we must talk about these anxieties we are beset by. Where did these anxieties come from? From the ancient Greeks, of course.

NATIONAL ARCHIVES PHOTOGRAPH

Relief from stress is easy in the modern world.

Chapter Six

The Birth of the New Science

Every day, in every way, things are getting better and better. Science has known this for a long time. Now you can, too.

☐ *What is the difference between "lemon taste" and "lemony taste?"*
☐ *How can you tell a rational number from an irrational one?*
☐ *Why couldn't a New Scientist like Quincy stay on the air?*
☐ *What is the last thing to go through an insect's mind before it smashes into your windshield?*

New Scientists at work.

THE MENTAL DEPARTMENT

Mankind has always had what science calls "that nagging feeling," even before he was fully aware that he was nothing more than a walking bag of leaky chromosomes. But despite the fact that the world has always been big and scary, early man's emotions were few. The dominant prehistoric emotion was fear (with its subsets, hunger and cold). We added more emotions as we needed them. The ancient Greeks gave us arrogance and bigotry, the Romans cruelty. The Dark Ages added gluttony, lust, and ambition to the list. The emotional toll remained constant until the Industrial Revolution, when we gained boredom, self-doubt, anger, and a false sense of security.

But as more emotions were created, our poor brains—like a baby's stomach—became upset. Unlike a baby, however, our brains had no Mommy to pat the pain away.

Thus the Victorian Age teemed with neurosis: the Romantic tradition and Impressionism, evolution and mysticism. All the young men were dark and Byronic; they wasted away at an early age, leaving behind beautiful corpses and slim volumes of morbid poetry. All the young women played the harpsichord and could faint at the drop of a hat. This was called "swooning." The Victorians did not have a life-style, they had what they called "a pitiable lot imbued with melancholy."

Humans were unhappy, and didn't know why. Today of course we know that electricity was the true reason, but it was a bold Viennese quack named Sigmund Freud who first put a name to unhappiness.

Freud

The ancients believed that ether was the pure air breathed by the gods. It resembled modern ozone. When the gods died in 200 B.C. the ether remained, to carry disease, scattered thoughts and ideas, even (as we have seen) light. It is from the word *ether* that we get the word *ethereal*, which is used to describe Tinker Bell, electronic music, and cornflakes.

Ethereal means "light" or "airy," an adjective not usually applied to Sigmund Freud. Freud sought his entire life to disprove the existence of not only ether, but logical thought itself. To some extent he succeeded.

It was Freud who internalized the ether and gave it a teutonic name: *subconscious*. Instead of messy demons, fairies, incubi, and succubi flitting around, he gave us the ego, id, and superego, each of which was responsible for a team of emotions: love, hate, guilt, remorse, grief, envy, and sex. When these teams were pulling together, you had a healthy person. When the emotions pulled at each other, you had what science calls a "normal" person.

If Freud was limited in his vision of our inscape (he ignored many important emotions such as indifference and cynicism, for example), he was still the first man to put a name to our anxieties. Of course one can say that name over and over again, and it doesn't erase anxiety. But even today,

FOR YOUR INFORMATION . . .

How Do You Tell a Psychologist from a Psychiatrist?

This is a real problem. The difficulty lies, as you know, in the fact that both psychiatrists and psychologists favor tweeds, horn-rimmed glasses, and soft, rubber-soled shoes. The men have beards; the women have their hair tied severely back. If *I'm* not sure, I use this simple test: Say to your suspect, "I hate my parents." The psychiatrist will say, "Why did you say that?" and the psychologist will say, "Thank you for sharing." As a rule of thumb, psychiatrists generally insist on being called "Doctor," while psychologists want to be called by their first name. A psychologist usually has just written a slim book about sex or success and hosts his or her own radio talk show. Psychiatrists spend their free time writing letters to the editor of the *New York Review of Books.* If all else fails, ask the subject directly, "Are you a psychiatrist?" The psychologist will say, sadly, "No, but I wish I was," while the psychiatrist will say, "Why did you ask that?" Neither psychiatrist nor psychologist is any fun at all at parties.

upper-middle-class people shell out hard-earned money to Freudians, just to have somebody to talk to. That's the kind of world we live in.

NEW SCIENCE CONQUERS MENTAL HEALTH

New Science can't erase that nameless dread that stalks our waking hours, but it can help us have more fun. New Science, at its best, is one big playground of diversions for grown-ups. For example, the exciting discoveries of the mid-20th century such as *power lunches, positive mental attitudes,* and *hair styling* allowed us to sculpt a self-image which in turn helped us to create what New Science calls a *lifestyle.* Then, New Science stepped up its activity. It changed the very foods we ate. By the late sixties, thanks to New Science, people began to take chemicals voluntarily into their bodies, in the hope of chemically mutating. Evolving. Having fun.

SO WHAT IS THE SUBJECT OF NEW SCIENCE?

Old science mastered the forces of gravity, light, work, and electricity. New Scientists know there are other, more crucial forces influencing the earth today, and New Science dedicates itself to discovering and controlling those forces to preserve the *lifestyle* we cherish. These forces are the most mysterious and powerful non-atomic forces in the world. More comic books and grade B movies have their plots hinge on these forces than any other. Any New Scientist worth his or her salt is constantly plagued by vague fears of impending doom. These forces can be used to protect ourselves from these fears. In our next chapter we will study these forces one force at a time.

☐ *What's the difference between "lemon taste" and "lemony taste?"*

■ There's no comparison. "Lemon taste" has to do with real or artificial products that simulate the taste of lemons. "Lemony taste" refers to nothing but to itself. It's a taste sensation that opens doors of perception that have remained closed since the dawn of time. It's an exit off the straight and narrow, a wild card in a stacked deck, a token that says "pass go, collect $200." Lemony taste is a hypothetical question that answers itself, a food additive so concentrated that one drop could change the world's eating habits and allow absolutely anything to become a food or foodlike substance. Look forward to experiencing "lemony taste" the next time you eat out. And when you do, don't thank the chef; thank science.

☐ *How can you tell a rational number from an irrational one?*

■ You can't, at least not without professional help. Playing psychiatrist to random numbers can only lead to heartache. I suppose you could try to start up a conversation with an unknown quantity, a general sort of dialogue about the weather or fashion. Avoid talking about sex, politics, or religion. If the number is truly irrational, such subjects will throw it into a blind rage, with possible disastrous consequences. Remember, inducing someone to break the laws of higher mathematics is a felony. No, if you suspect a number of being irrational, consult a trained mathematician or a fourth-generation computer. Otherwise, you're courting disaster in a big, big way. Like a zillion. That's a big number. Isn't it?

☐ *If Dr. Quincy could interpret nuclear magnetic resonance strata and mass spectra better than his clinical chemists, why couldn't he get ratings good enough to stay on the air?*

■ TV shows pool their knowledge to help each other out. Quincy gained his astounding medical knowledge from the episode on "Star Trek" in which Spock's brain was stolen. As you recall, Dr. McCoy hooked himself to an intelligence booster to gain the knowhow he needed to put Spock's brain back in place. Quincy borrowed this machine, and his show went off the air when the Nielsen families borrowed this machine as well. TV science, unfortunately, is just science with a shelf life. When you see that commercial that begins, "I'm not a doctor, but I play one on television. Many people play doctor at home," he is giving himself thirty seconds worth of medical knowledge. If you were an actor, in other words, you could be a doctor, too.

□ *What is the last thing to go through an insect's mind before it smashes into your car windshield?*

■ By a remarkable coincidence, science has recently shed light on this problem. After a long and diligent search, science located a number of psychic insects. We duct-taped these telepathic six-legged creatures to the hood of a '62 Studebaker. Then we drove through Wyoming in August at 100 miles an hour. Tiny instruments picked up the final thoughts of so-called "normal" insects as they smashed into the windshield. Twenty-three percent of these final thoughts were "Uh-oh." Another twenty-five percent were "Oops," followed by fifteen percent that were simply, "Car!" The remainder of these thoughts were a hodgepodge including, "Hungry," "Tired," "Woh! Big and Shiny," and "Hey, watch where you're going."

NATIONAL ARCHIVES PHOTOGRAPH

This young lady has not yet learned: DNA is not a plaything.

Chapter Seven

Natural Forces that Must Be Tamed by Science

> Nature is the enemy. Only by embracing technology can we resist the destructive lure of entropy.

☐ *Does virgin sacrifice have an effect on volcanic eruption?*
☐ *Why does the sensation of being aboard ship stay with us long after coming ashore?*
☐ *Is there such a thing as an unforgivable sin?*
☐ *Why can't you divide by zero?*
☐ *Where do the original voices go when a foreign film gets dubbed?*
☐ *Why do tongues stick to ice cube trays?*
☐ *Why isn't a telephone conversation backward for one of the callers?*
☐ *What is the "fifth force" of nature?*

YOU'RE JUST A PAIN IN THE NECK!
HUH! YOU'RE NO HEADACHE POWDER YOURSELF!

NATURAL DEPRAVITY

As we have seen, the Victorians knew what Modern Joe and Jane can't remember. That a man left alone will reach "bottom."

The power of natural depravity is so great that a normally wholesome individual will, if given the chance, become lower than an animal. In a rush of hormones and endorphins he will lose his IQ.

Once depraved, such a man is incapable of recovery. He will always need professional help. Here you will see the obvious advantage of being a New Scientist: when you become depraved you can turn to yourself. No high fees. No frustrating phone calls trying to make an appointment. You are doctor and patient in one neat package.

ATTENTION SPAN

Not the mighty bridge that links land masses over bodies of water, no, the attention span is a lowly thing, capable only of bringing thoughts from one synapse to another. Though small, this force in many ways is the most important force to be controlled by the New Scientist. At this point you are still a lay person. You may or may not have noticed that the chapters in this book have become shorter. If you *did* notice, you have the makings of a New Scientist. Your attention spans will become longer and longer. You will begin to use large ungainly words to express easily understood concepts. If you did *not* notice, there is probably no hope for you. You should probably stop reading this book and turn on the television.

THE POWER OF SUGGESTION

The power of suggestion operates through compulsion. Obsessive/compulsive behavior is not a natural force, but rather a tool used by natural forces to increase human misery. Compulsion is the voice that screams in your ear, "Hurry! Sale ends tomorrow!" when you had no prior interest in buying anything. And yet you will buy something on sale that you neither wanted nor needed, and the subsequent egg on your face will be put there by compulsion.

Waiters employ (use) the power of suggestion when they bark "Enjoy!" after serving you. Sales clerks misuse the power of suggestion by chirping "Have a nice day!" after they take your money.

Scientists use the power of suggestion through catchy phrases like "better living through chemistry," and "smart missiles." Like earwigs, these phrases slowly but surely crawl into the brain. Once inside, you can't get them out.

REMOTE CONTROL

It would be a foolish person indeed who did not respect and admire the incredible power of remote control. Science respects remote control, as much as science respects anything. Nowadays, in order to operate any device by remote control, you must pass a citizenship and mental health test,

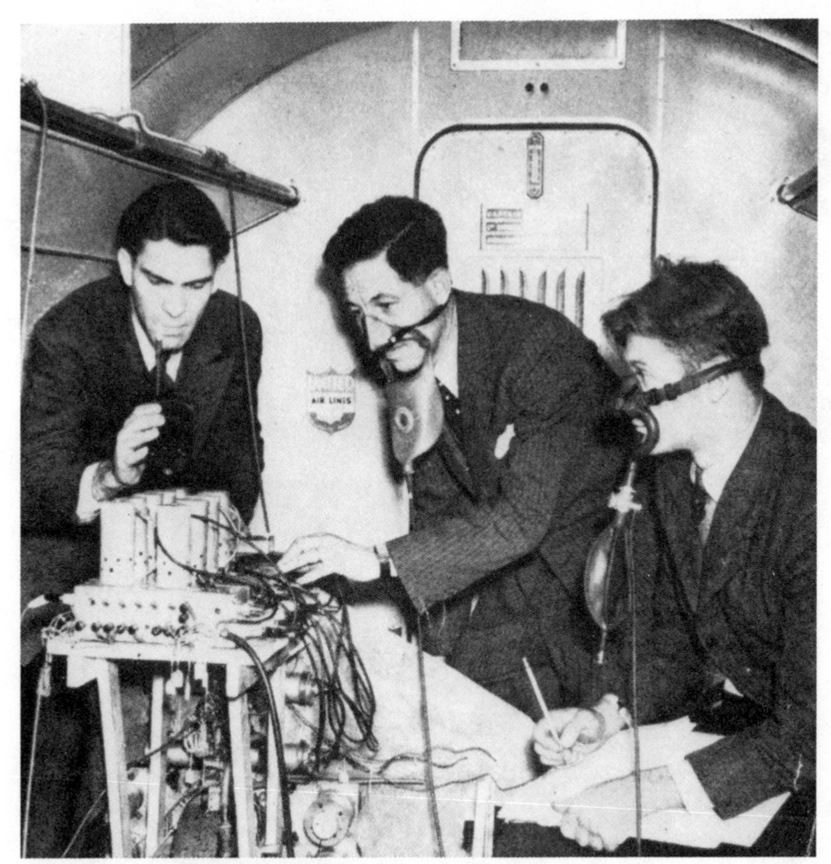

Harnessing the hidden forces.

administered by urinanalysts and polygraph operators employed by the federal government. If you fail this test, you could be jailed for life. No, the government does not take remote control lightly, and neither should you.

HYPNOTISM

When someone smells a flower and makes an ugly face, chances are that person is hypnotized. In fact, hypnotism is classified as a disease by the American Florists' Association. In their hatred of hypnotism, they are not alone.

Hypnotism was first discovered by Franz Mesmer, the

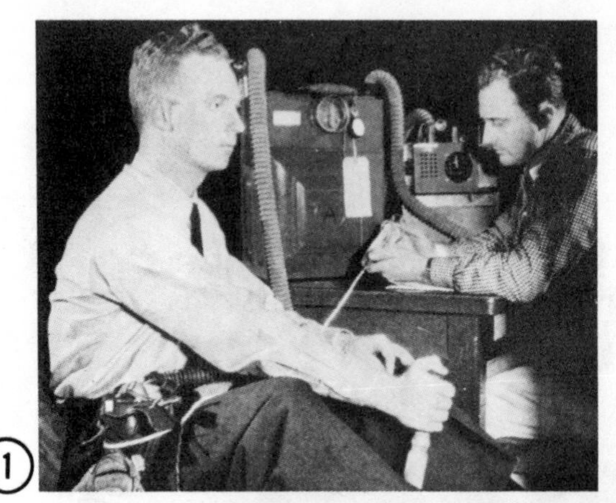

Test subject undergoing hypnosis.

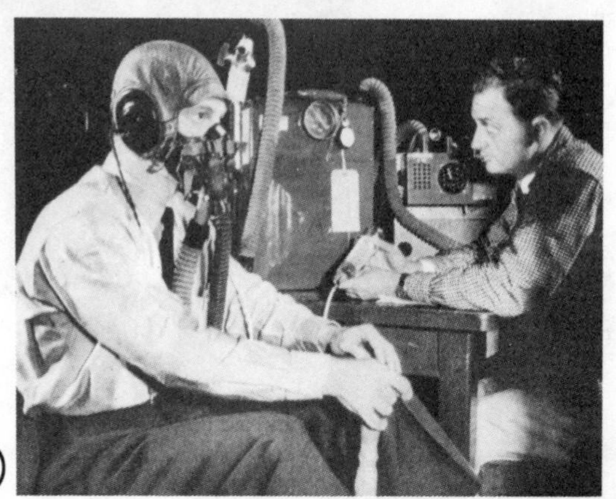

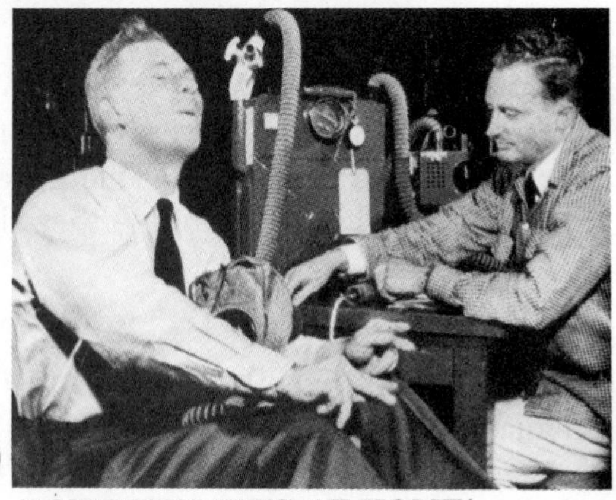

DON'T TRY THIS AT HOME!

Hungarian philosopher who also invented yogurt. He had been trying for some time to persuade his wife to stop rearranging the furniture in their house. It seems she had a compulsion to redecorate their home every week; Mesmer sought to rid her of this compulsion by spending half an hour every night talking to her while she was asleep. We have no record of whether she became satisfied with the arrangement of their furniture, but Mesmer did note that she began to complain that flowers stank and candy tasted sour.

Hypnotism is still quite the rage among adolescent males and scientists who hope to attract women who otherwise wouldn't be interested in them. I tried it myself, using a metronome and a bottle of cheap champagne. I still have a scar from that metronome, and she took the champagne with her when she stalked out of my rooms. Yes, hypnotism can be a powerful tool for stimulating the imagination. It should be cautioned, however, that one must have a will before one can bend others to it.

PERPETUAL MOTION

Perpetual motion is impossible. Each of the laws of thermodynamics makes this quite clear. Yet perpetual motion machines continue to be invented, because many inventors are unaware of the laws of thermodynamics. So if you plan to build a perpetual motion machine in the near future, I would advise you to forget the laws of thermodynamics. If you don't know the laws of thermodynamics, ignore this section.

UGLY WALLPAPER

Ugly wallpaper can kill. Many an unsuspecting traveler has checked into a hotel room only to wake the next morning the shell of the person he or she was the night before. The culprit? Ugly wallpaper.

TO DO AT HOME

THE OWSLEY SANDOZ LSD EXPERIMENT

This was the first attempt to prove that light is composed of both particles and waves. As the Tibetan Book of the Dead tells us, light and LSD are really the same substance in two different forms. It's up to us non-Tibetan, undead types to prove the scientific verity of this ancient truth.

Of course when one is under the influence of LSD it's hard to prove anything. Especially to the satisfaction of the Scientific Community.

Since I don't suggest that you take LSD, we'll do the next best thing. We'll simulate the altered state by altering our laboratory.

First of all, we'll paint the walls lime green, canary yellow, and hot pink. Great, nauseating swirling patterns of color. Next we'll install a black light in the center of the room and then place several fluorescent posters on the walls. Then we'll burn large amounts of strawberry incense, continually.

Now, we'll put on the appropriate music. The Doors, Jimi Hendrix, Iron Butterfly (many important discoveries have been made during the endless drum solo in "Inna Gadda Da Vida"). And lastly, we'll ingest something that will make us more susceptible to discovery.

It is not the purpose of this book to prescribe for your psyches, so no attempt will be made to suggest what substances should be ingested. But, gentle reader, this

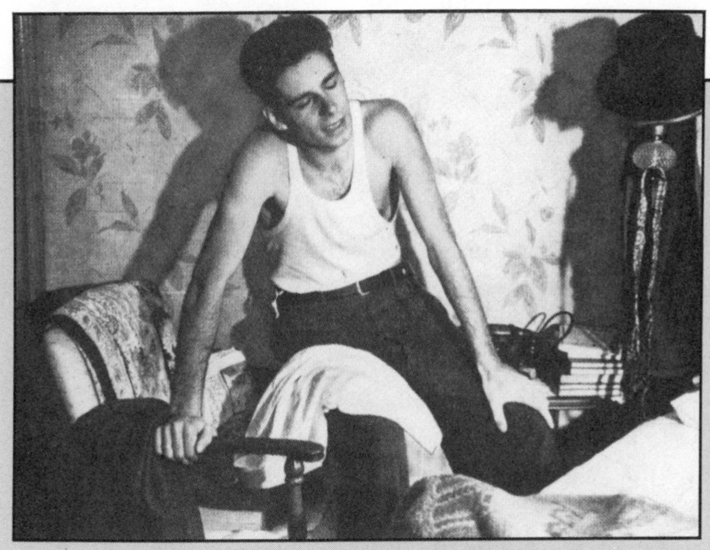

Budding scientists often have a hard time.

scientist cannot speak highly enough of coffee. Caffeine is a psychotropic drug that packs a wallop and is relatively inexpensive, considering what it does. The most powerful and expensive form of coffee comes from the Ethiopian Harararrara bean, the largest, blackest, and most pungent bean of all.

Once we have ingested several cups, then we are ready for some intense experimentation. We turn all ambient lighting down, allowing the black light to elicit its magic glow from the posters. Look at our fingernails, they're glowing pale blue! Trippy! Is the music loud enough? We can always turn it up. It's our right as scientists!

Now we notice a flashing that happens in the corners of our eyes, on the edges of our field of vision. This is what has been described as "acute caffeinitis." Sometimes the flashing appears to be related to the music, sometimes not. This may be proof of the particle/wave nature of light. It may be an indication of synaptic failure due to sensory overload. We don't know.

As in any experiment, one should allow enough time for the variables to vary, the phenomena to phenomenize. But eventually we will know all we need to know about ourselves, caffeine, and Iron Butterfly. As to the exact nature of light, it's hard to care really when you've drunk that much coffee.

Some ugly wallpaper doesn't look so bad during the day, but at night, when you can't sleep and you're watching the car headlights march across the ceiling, that's the time ugly wallpaper gains its power.

Grotesque faces, strange bloblike forms, undulating streams of dismembered body parts—these are the masks ugly wallpaper wears at night. These masks have the power to kill the spirit, to tear the psyche from its snug house of reason and then burn that house down.

Nobody can stop you from staring at ugly wallpaper if that's what you want to do. But the responsible scientist strongly suggests that ugly wallpaper be eliminated from hotel rooms, hospital solariums, and dentists' office waiting rooms.

The voice science uses to suggest this course of action is not the harsh voice of compulsion. It is not even the tugging voice of intuition. No, science uses technical writing such as this to make its point. Clearly. Unmistakably. But above all, gently.

IS IT PAINFUL TO MASTER THESE FORCES?

Yes, of course. Thomas Aquinas proved conclusively in 1004 that in order for anything to be worthwhile it must be painful and take much longer than anyone expected. His treatise, which translates from the Latin as "Them's the Breaks," is universally accepted as the document that paved the way for New Science. Certainly, science might have been able to accomplish some measure of progress without Tom Aquinas's help, but science would have been bland and colorless. Its biggest moments would have concerned plant genetics. Crop research.

Thanks to Tom Aquinas, scientists know pain. And the rest of us do, too. Scientists thus eagerly await their role in the Battle of Armageddon, in which they expect to receive at last the attention and respect they deserve.

☐ *Does virgin sacrifice have an effect on volcanic eruption?*

■ Yes. A woeful lack of virgins in today's world, combined with a pitiful lack of nerve on the part of geologists, is the main cause of the Mt. St. Helens eruption. Forget all the talk of tectonic plates and volcanic pressure. The truth is, ancient gods are asleep in there, hibernating as it were, and when these gods wake up, they wake up mean and hungry. Any hungry creature deserves a snack. If we don't have the necessary virgins willing to make this sacrifice, perhaps a couple of Quarter Pounders would curb the big god's appetite. It'd cost a few bucks, sure, but isn't saving a mountain worth it? Geologists, take note.

□ *Why do I still feel the sensation of being aboard a ship a long time after coming ashore?*

■ The sensation you describe, which science calls a "memory blooper," is the same phenomenon that causes you to feel like you have your hat on after you've taken it off and like you have it off when in fact it's on. There is no cure for this synaptic short circuit. Early blind tests reported some success with increased intake of a mixture of zinc, riboflavin, and 2.2.4. trimethylpentane. These tests were stopped when several subjects began to believe they were wearing a hat aboard a sinking ship, when in fact they were under heavy sedation in a sterile white room. Scientists do this sort of thing all the time—place students on work-study in a sterile room, ask them to remove their clothes, and then watch them through a two-way mirror. Science will continue to do this as long as students are willing to do this. Hope I've answered your question.

□ *Is there such a thing as an unforgivable sin? Is so, what is it and how do I commit it?*

■ The worst sin, the act that all races and religions agree is against the laws of nature, is to attempt to divide by zero. Even computers will stop, horrified, in their tracks and chastise any operator crass enough to attempt such a thing. Division by zero may leave an indelible stain on the soul, which can be erased only by a full confession to a qualified mathematician. Even then, this sin must be atoned for in a purgatory filled with unbalanced checkbooks and tedious exercises in long division. But that's the way it has to be; otherwise, the entire system of checks and balances would fly out the window.

□ *Why can't you divide by zero?*

■ I can and often do divide by zero, but only after I've made the necessary preparations. First of all, I fast for forty-eight hours, consuming during that time only mildly fluoridated water. Next I don my special Mylar/Teflon division-by-zero suit. Then I put on a digitally recorded

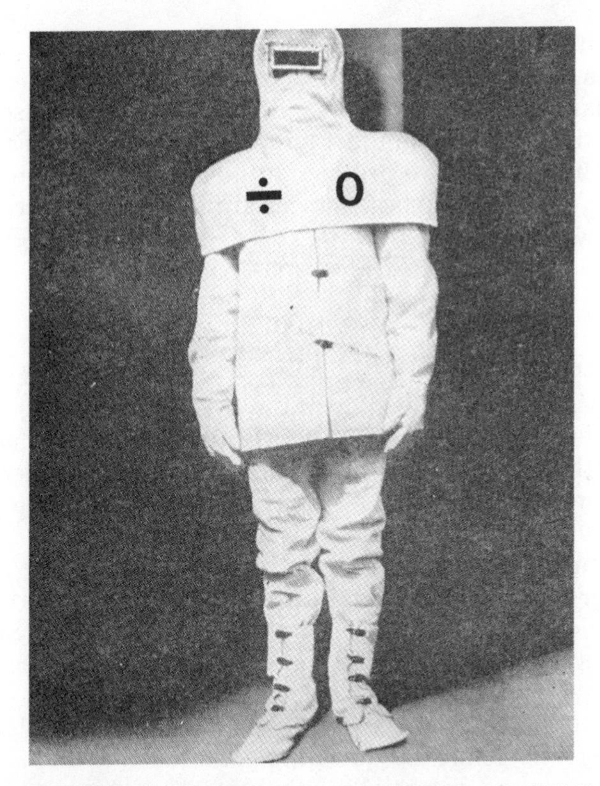

compact disc of Gregorian chants and begin with dividing very small numbers by other very small numbers. As the numbers get smaller, the sparks begin to fly. If all goes well, I take a deep breath and divide a very small number by zero. There's a flash of light, a muffled roar, and when I come to, the lab is filled with smoke and the scent of burning Mylar. So you see, you can divide by zero if you really want to. But chances are you just don't want to badly enough.

□ *I am a foreign film addict. When you watch a foreign movie that has been dubbed, where do the original voices go?*

■ They don't go anywhere. As you know, when you wake up foreigners in the middle of the night, they talk normal. So essentially there are two versions of any foreign movie: one released in foreign gibberish and one shot in the middle of the night after the foreigners have been rudely awakened. As a sop to liberals and people who think a thing has an intrinsic value because it's French, the foreign-language version is the one that's usually released, with subtitles—so you can feel like you're reading at the movies, thus killing two pretentious birds with one stone. Don't be fooled. All of these movies are shot in New Jersey. Nobody really talks like that. It's all a sham, and foreign movie fans are just victims of a massive con game.

□ *Why does my tongue stick to the ice cube tray?*

■ Next time you put the tray in the freezer, try filling it with water instead of your tongue. I think you'll find that your parties will be more fun, your drinks will be colder, and you won't have to talk to your guests in a voice like Elmer Fudd's. As to the why of it, under subzero temperatures, your tongue secretes a clear paste that bonds like superglue. Superglue was originally created by forcing small children to put their tongues on cold swing sets. So remember, dress your tongue warmly. If you must put your tongue where it doesn't belong, make it a postage stamp. And never ever kiss if the temperature falls below thirty degrees. You're only opening yourself up to an embarrassing and potentially lethal social situation.

□ *Why isn't a telephone conversation backward for one of the callers?*

■ The words are backward for both callers, but in the audio equivalent of an optical illusion, the ear automatically adjusts. The telephone operates the same way as a two-way mirror. Your image or voice is reversed on the surface or speaker, but the person on the other side sees or hears you normally. So any telephone company is, in effect, a hall of mirrors. With all those voices it's easy to become confused. The dial tone was invented as a kind of mantra to soothe you in the cacophony. It is only the dial tone that keeps telephone operators from becoming hopelessly insane. It was also for health reasons, not financial, that the federal government broke up AT&T. And now, with answering machines and cordless cellular phones, we face a greater threat to our sanity than we have ever known. I have tried to speak to the surgeon general about this, but his line is always busy.

☐ *Could you please explain the newly discovered "fifth force" of nature?*

■ Glad to help out. The newly discovered fifth force is called *force of habit,* without which none of the other forces would work. Why do men and women attempt to live together despite ever-rising divorce statistics? Force of habit. Iron is attracted to magnets because it's lazy. Atoms stay together because they have nothing better to do.

When Galileo dropped a pound of feathers and a pound of lead from the Eiffel Tower, he wasn't proving anything, really; he was just bored. If today some scientists say Galileo is wrong and a feather drops faster than a rock, they're just saying that because they have to say something. If they don't say something, they lose their research grants. Then they'd be unemployed and even more bored.

TIME TO MEET THE MASTER

There is one man who can epitomize the proper use of all natural forces, one man who can be a role model for the rest of us, one man who can encourage others in the correct use of the higher powers. The time has come to meet that man.

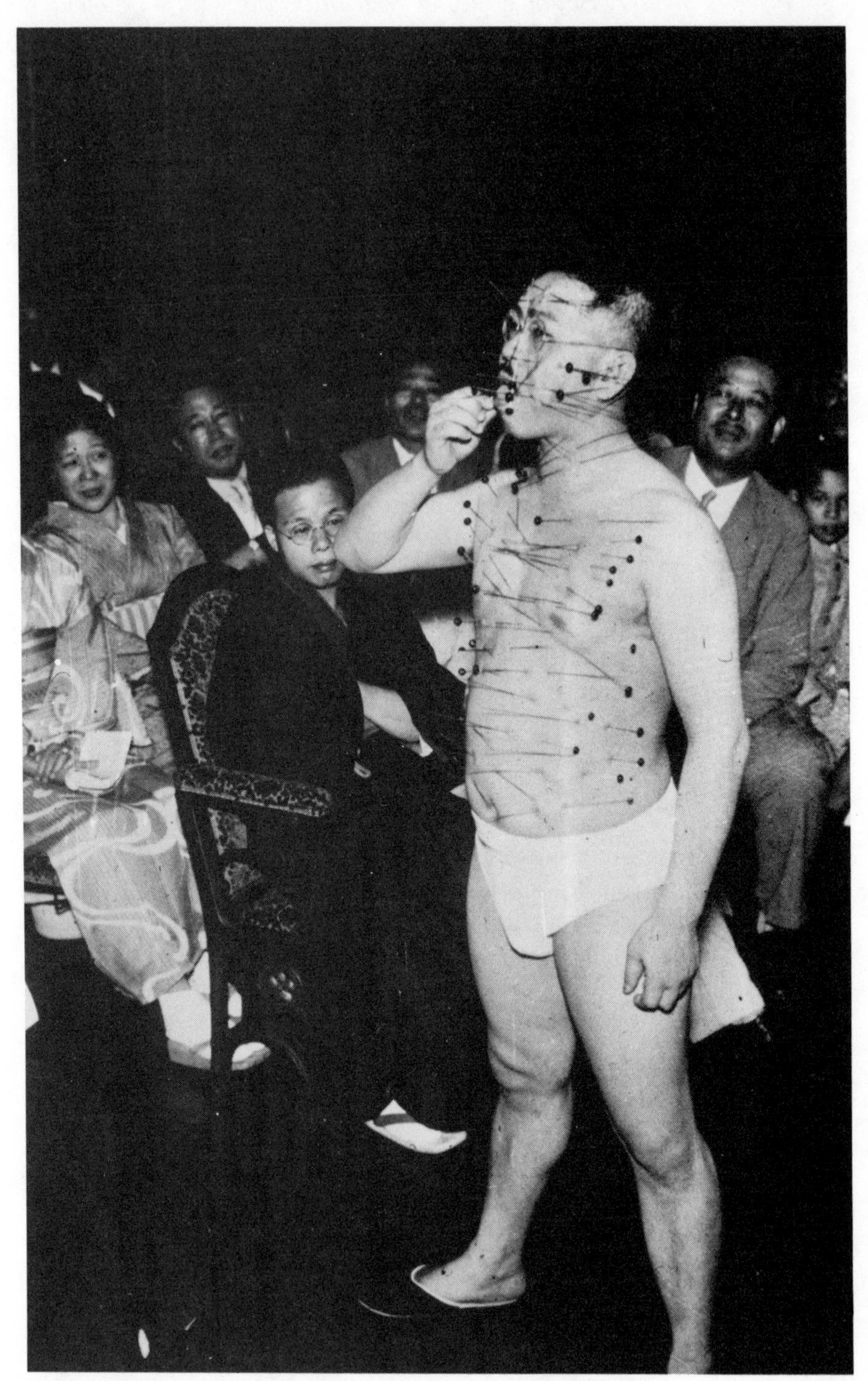

NATIONAL ARCHIVES PHOTOGRAPH

Mastery of pain is essential for the New Scientist.

Chapter Eight

The Model of the New Scientist

Does the Scientist of the future exist today? We think so. Read this chapter and you'll think so, too.

☐ *What happened to all the eight-track tapes?*
☐ *What does Dr. Science do with all the letters he doesn't use on his famous radio program?*
☐ *How does the male sperm get to the female egg (in humans)?*

The Fortress of Arrogance.

DR. SCIENCE AS ROLE MODEL

For Dr. Science, science is not a mere collection of facts, but a rebellious way of life! It is the plant that grows from the spot his bitter tears have watered. Oh, Dr. Science knows all too well the chief exports of Bolivia. Yes, thank you, he can name the seven deadly sins, the three divisions of ancient Gaul, the four reasons to go on living.

And today he is a man who knows what he knows. A man who is not ashamed to say he knows what he knows. Certainly, there are some things he didn't learn. Human intercourse. How to make friends. But these things didn't interest him as a child, and they don't interest him today.

He considers himself fortunate to have had the opportunity to look at everything twice. Old Science was the thesis. And the views he now holds are the antithesis of Old Science: New Science. With this chapter we forge away ahead to the third step: synthesis (or prosthesis).

But before we take that step, we must take a look at my, I mean Dr. Science's, formative years.

A Life in Science

Science was Dr. Science's main concern for as long as he can remember. He was weaned on a test tube filled with condensed milk kept at a constant sixty-four degrees Centigrade. His bassinet was a prototype for a perpetual motion machine. It would rock him back and forth for days on frictionless Dynel bearings. He soon grew accustomed to its incessant rocking, realizing even in his infant brain that there were natural forces at work that he was powerless to control.

By the age of three he knew his logarithms and most of the geometric functions by heart. By the age of four he had deliberately forgotten them. On his fifth birthday he published a proof for generating and verifying random numbers.

Unfortunately, this prodigy did not endear himself to his teachers, and he was obligated to repeat every grade from kindergarten to high school. Now that Dr. Science looks back on this, he can dismiss some of his resentment of the "authorities" who helped make his childhood the living hell that it was.

THE GENIUS COMES OF AGE

Like most boys, Dr. Science joined the Boy Scouts for a time. He dabbled at earning merit badges until he was asked to leave. Reasons? Something about his extreme arrogance. His know-it-all attitude. But Dr. Science would not pretend not to know things he knew just to make some den father feel important.

By the time he arrived in high school, he knew his destiny. When the guidance counselor asked him to consider taking nonscience courses, he laughed his most scornful laugh and strode out of the office. This lack of cooperation meant that it took him six years to receive his high school diploma. This only increased his contempt for academia.

THUMBNAIL SKETCHES OF THE GREAT

Gustave Flaubert—Inventor and Novelist

The novel **Madame Bovary** was the creation of an inventive person who didn't stop creating when he put down paper and pen. Gustave Flaubert was also an inventor, today best remembered as the inventor of Flubber, a product so amazing that it inspired Walt Disney to make a movie starring the magical, amorphous blob.

Flaubert was a great novelist and inventor, but a bad businessman. His failure to obtain American patents for Flubber cost him millions in lost licensing fees. Disney Studios delighted the masses with the zany antics Fred McMurray enjoyed with the novelist's creation, but a hundred years before in France, Flaubert became morbidly depressed and impoverished.

In 1880 the end finally came. He chose the same fate as Madame Bovary, his creation. His ashes were carried in a bronze catafalque on a burnished barge down the mysterious Nile, where he finally found a peaceful resting place among the proud pharaohs, a peace he never found in France.

If Flaubert were alive today: He would be a hairdresser and part-time travel agent, living in Florida.

His university days were much gentler on his psyche. Perhaps it was his advanced age. Starting his freshman year at the age of thirty gave him a self-confidence that allowed him to nurture his arrogance. He cultivated the air of a relaxed gentleman scholar. He began to smoke a pipe. He wore tweed jackets with leather patches on the elbow.

THE MASTER'S DEGREE

Although Dr. Science's grades were poor, after graduation he was able to talk himself into graduate school at a university that wishes to remain anonymous. He was to remain at Institute X for the next five years. Personality conflicts with the faculty and a perfectionist streak in his work (which demanded that he constantly and compulsively check and recheck his data) finally persuaded the faculty to award him the degree of Master of Science. This was an important day in his life.

The postman must have noticed the tears in his eyes as Dr. Science brushed the junk mail from the nine-by-twelve envelope containing the document that certified him as an expert. Although he was alone in his furnished room, he quickly uncorked a bottle of the finest vintage cognac.

When he came to, paramedics were shining flashlights in his eyes and asking him if he knew his name. His proud reply has been preserved for posterity: "Not only do I know my name, sir, but I know more than you do. For I have a Master's Degree, in Science!"

Today these words are etched in stone in the arch that marks the entrance to the Fortress of Arrogance, the home of Dr. Science, as well as the laboratory where he originates his instructional radio program.

DR. SCIENCE AT HOME

Dr. Science's home is an unassuming two-story brick house, built in the prairie modern style. In it he resides with his

Skeptics and mockers do not disturb New Science.

pets, a blind Yorkshire terrier named Oscar and two short-hair Persian cats, Proton and Neutron.

Musical instruments abound, including a harpsichord that dominates the living room, and a large collection of flasks filled with varying amounts of water, which Dr. Science calls his "water harmonica." These he rubs with damp fingers, producing an eerie ringing noise.

His bedroom is the smallest room in the house, not much bigger than a closet. In it there is a cot and a bust of the scientist and philosopher, Leibnitz. A lone strand of Christmas tree bulbs provides the only illumination, save for a dust-covered window that seems to open onto a brick wall.

The largest room of the house is the bathroom. There, a bathtub the size of a small swimming pool lies perpetually filled with hot mineral water. The stench of rotten eggs fills the rooms. None of Dr. Science's pets will venture into the bathroom; instead they stand outside the door and howl, especially when Dr. Science is in the tub.

THUMBNAIL SKETCHES OF THE GREAT

Luther Burbank—The Gardner Who Wouldn't Take No for an Answer

It seemed the rain would never stop, that April Day in 1925. Santa Rosa, California, was a town that expected rainy winters, but the winter had lasted long into spring, and the farmers were impatient to get on with their planting. Luther Burbank furrowed his heavy brow and thought, "If only there were some way to produce a plant that didn't care a hoot about weather."

Of course this was just a dream, one of the many possible dreams of a man stuck indoors on a rainy day. A normal man would have lit his pipe and gone on dreaming. Since Luther Burbank didn't smoke, he decided to go into his basement laboratory and turn his dream into reality.

All that night Luther Burbank worked, and dawn found him groggy but happy. On his workbench lay the first avocado the world had ever seen. A hybrid between a pear and a can of motor oil, Burbank's avocado would change the face and taste of Mexican cooking.

Of course no prophet is honored in his hometown. Burbank died, penniless and virtually unknown, in a bowling alley gutter in downtown Santa Rosa. Today this bowling alley is an office building, and the elevator of that building proudly bears a plaque: "The Luther Berbanc (sic) Memorial Elevator. Maximum Load, 4 Adults."

If Luther Burbank were alive today: He would be the leader of a Boy Scout troop and live somewhere in northern California.

Since Dr. Science has supplanted his need for food with daily vitamin injections, he has no need of a kitchen. This room has been transformed into a recording studio, from whence the "Ask Dr. Science" show is broadcast all over the free world.

Dr. Science has his own personal microphone, which is the largest in the world. It is roughly the size of a refrigerator. His research assistant's microphone is the smallest in the world. It is so tiny it has actually been surgically implanted in the assistant's throat. This is to ensure that Rodney will not lose his microphone, because it is very expensive.

The "Ask Dr. Science" show is recorded on a special German tape recorder, using giant spools of titanium wire. These are then delivered by unmarked cars in the dead of night to the radio station, where they are beamed by satellite to avid listeners everywhere.

Dr. Science's backyard is completely given over to hydrangea bushes. These emit an overpowering cloying odor, one that the sensitive nose finds pleasantly sweet at first, then sickening.

The garage is a storage shed, as Dr. Science will not and cannot drive. He walks the thirty blocks to his lab, six days a week, fifty weeks a year. He spends one day a week in silent retreat at the Fortress of Arrogance; and two weeks out of the year are spent either in his massive bathtub or in Las Vegas, where he gambles compulsively for his entire stay.

SO WHAT DOES THE FUTURE HOLD IN STORE AGAIN?

A lot of wishful-thinking no-brain humanists who can't tell fecal matter from shoe polish view the future with alarm. But perhaps that's expecting too much from the future. What can we expect, realistically, from New Science? Let's take Dr. Science's work as an example.

Dr. Science no longer eats, bathes, or cohabits with the opposite sex. When New Science is through with us, America will live much as Dr. Science does. People will replace their reliance on food with daily vitamin injections. Folks will replace their hair and skin with look-alike substitutes made of the miracle fiber Dynel. Men and women will finally realize they're better off just leaving each other alone.

If this sounds a bit grim, it's because you're still looking at the world through the eyes of Old Science. Think of Old Science as a pair of scratched spectacles and New Science as a pair of extended-wear contact lenses. Now the picture becomes clear, doesn't it? By following the rules of simple hygiene, and through regular checkups at the specialist of our choice, we will enjoy a new level of vision, approximating what our primeval (before 1880) ancestors saw without the aid of glasses.

□ *What happened to all the eight-track tapes?*

■ I still have mine. What did you do with yours? Science can't help you take care of your personal belongings. It can, however, suggest new uses for old technology. I've used my eight-track tape machine to manufacture durable and stylish clothing. I simply take my old socks and insert them into the slot, followed by any old eight-track cartridge. Then I let the machine run through its cycle. When it's played all eight tracks of distorted music, I withdraw the cartridge and usually find a new shirt or pair of trousers. It's an example of the kind of thing you turn up through experimentation.

□ *What do you do with all the letters you don't use on your famous radio program?*

■ That depends. The majority are used for wallpaper in the master bedroom, where I spend many a sleepless night gazing at the walls and wrestling with brainteasers. Those letters and cards, which include pictures of kids, wives, and husbands, are made into a cheery display that goes up on the mantle. Some letters are used to create nuclear models in the lab, with the help of glue and a bit of string. The rude questions and those questions that man dare not ask—these I destroy in a special ceremony, the details of which need not concern you. The questions written on sturdy bond with a high rag content are saved. I white out your questions and use the paper as my personal stationery.

□ *In order for people to have babies, the female egg has to be fertilized by the male sperm. How does the sperm get to the egg?*

■ It hitchhikes. There are small arteries (or highways if you will) in the man's urethra. Small foreign bodies (or cars if you will) pull over and pick up these little *spermatozoans* (or tiny hitchhikers), then drive over the speed limit as fast as possible before the natural acids in a woman's body (the highway patrol if you will) pull the sperm over for reckless driving. Once they get to the ovum (or garage) they get out of the car, turn out the lights, lock up, take off their shoes, and watch television until they fall asleep. This is where babies come from: small suburban ranch-style homes hidden deep in a woman's body. I hope I've answered your question. Good luck on your date tonight.

WHAT WE'VE LEARNED SO FAR

New Science has thrown Old Science out on the dung heap. Dr. Science is the epitome, the very model, of the New Scientist. Dr. Science is a genius. Dr. Science is a hero for youths everywhere. There are many forces in the universe we should not abuse. Dr. Science loves all of you, with as much love as he is capable of having, which isn't much, but is better than nothing.

REVIEW QUIZ

1. Happiness is (a) a thing called Joe. (b) Nonexistent. (c) possible only for a New Scientist. (d) a warm puppy.
2. Were any ancient Greeks discussed in this chapter? Discuss.
3. (True or False) Truth is in the eye of the beholder.
4. Is microwave popcorn an example of Old Science or New Science? Discuss.
5. Would you vote "yes" on a bond issue to have a major freeway in your community renamed "Dr. Science Expressway?" If you answer "no," do not bother to finish this book. If you answer "yes," write your city council immediately. Better yet, give them a call.

NATIONAL ARCHIVES PHOTOGRAPH

Build a human radio at home!

Part Three

Do It Yourself: You Can Become a New Scientist

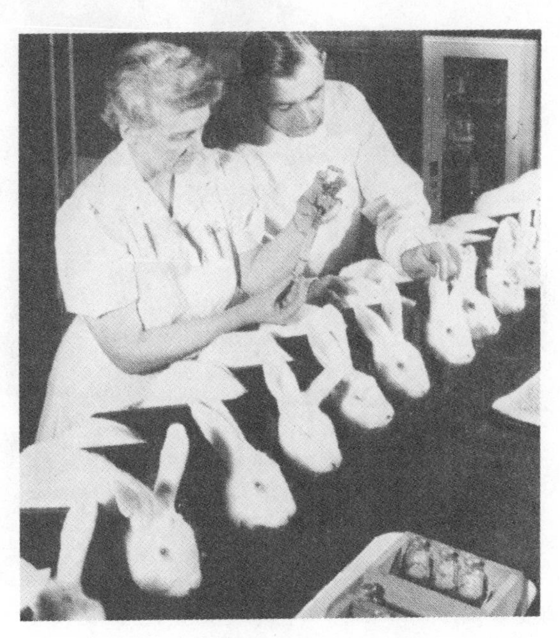

NATIONAL ARCHIVES PHOTOGRAPH

When Science comes knocking at your door, check before you answer.

Chapter Nine

The Wonders Around Us

Personal transformation is easy, if you follow a few simple rules. Do yourself a favor, follow directions for once.

- ☐ *Why do scientists drink so much coffee?*
- ☐ *What does FDIC stand for?*
- ☐ *Where exactly is the scientific community? Can* anyone *live there?*
- ☐ *Why do scientists in the movies always have a foreign accent?*

NATIONAL ARCHIVES PHOTOGRAPH

Modern telecommunications offer hands-free convenience!

NEW SCIENCE HAS THE ANSWER

When little Johnny wakes up, it's to modern electronic music instead of an annoying electric buzzer. A hot and powerful shower plasters his hair to his head, which he can then whip into shape with nongreasy and hygienic styling mousse. His breakfast is waiting for him, freshly agitated from the microwave oven. Then Dad drives him to school in a compact plastic foreign car that gets over thirty miles to the gallon. And little Johnny spends the rest of his day learning about computers, the wonders of the insect world, the mystery of the dinosaur, the lightning speed of pocket calculators. After school he is beaten severely by a youth gang, and he is taken to the emergency room at the local hospital, where the wonders of modern medicine make stitches painless and fun!

Yes, it seems that every time you turn around there's some new scientific breakthrough. Opinion polls tell you what you're thinking before you've even thought. Amazing new snacks combine crispness and chewiness in combinations thought impossible as recently as 1956. Why then, do these incredible breakthroughs make you nervous?

Because you are not a scientist. This is a simple answer to a complicated question, but what does it mean? For one thing, it means that scientists are seldom nervous, and when they are they do something about it. They discover a new element, suffocate a small animal, or take heavy medication. They apply for a research grant, destroy life as we know it, or develop an exciting new method of nuclear waste disposal. The point is they keep themselves busy.

In this chapter you will learn how you can become a scientist in your spare time. You will learn that life is indeed, as Sonny Liston put it, "a funny thing." But you will learn that it's not funny-ha-ha, but funny-peculiar.

So before we begin the painful process of self-realization that will lead to science, we must list the pros and cons of science. Study this list carefully and make up your own mind. If you choose to stop the process after examining this list, that's up to you. But consider carefully. Once begun, science is like the cold war. There's no easy way to stop.

THE PROS AND CONS OF BECOMING A SCIENTIST

You Get to Wear Special Clothing

Pro: Lab coats come in only one color. White. The New Scientist doesn't have to spend hours agonizing over which clothes to wear, and can save the decision-making brain cells for hard laboratory work. Lab clothing *must* be white; it's an unwritten law. Anything off-white or pastel looks like an artist's smock. While you were putting in a lot of hard years in college wrestling with logarithms, the arty types were goofing off. Now you want to be completely sure you are never confused with "them."

Con: A scientist's smock is no different from the smock of a medical doctor. The danger of being lumped in with the

Even Mom can be a New Scientist!

"health professions" is great. I suppose physicians have their place, but I wouldn't let one near *my* cyclotron.

People Assume You're Smarter than You Really Are

Pro: When you are a working scientist, people are afraid to attack you in the intelligence department, even though you probably stopped growing intellectually at age 19. You learned in graduate school how to speak in a way that attacked others where you yourself were weakest. Now you have constructed a virtually impregnable fortress for your ignorance. Congratulations!

Con: The price you pay for intellectual stagnation is boredom. As a tenured scientist, you can spend the rest of your life sneering at those colleagues whose published opinions oppose the thesis you wrote twenty years ago. Is that *your* idea of a good time?

Without modern science, life would hold little pleasure for these boys.

You Get to Spend Government Money

Pro: The government, despite what you read in the papers, has lots of money. The greatest chunk of this money goes to Defense. Science has made itself very close to Defense. This is an understandable motivation on science's part from the money point of view. Scientists who don't mind hanging out with military types benefit the most from government windfalls.

Con: You might be despised by poor and bitter scientists who don't receive government money. Of course most of those scientists are in the field of medicine, which, as a field of science, is almost beneath contempt.

Also, you may feel that the day-to-day hassle of dealing with the government just isn't worth it. You may feel that unless you are Secretary of Defense, you won't be given free rein to indulge your scientific fantasies.

You Get to Play with Toys that Are too Expensive for One Person to Own

Pro: Much of this wonderful equipment is in the hands of universities. Universities, strictly speaking, are not part of the government or the department of defense. So a Con for the previous statement could be a Pro for this one, scientifically speaking. In either case, you can horse around with multibillion dollar cyclotrons, linear accelerators, and make really big sparks at all hours of the night.
Con: There is something in the university that makes you feel too safe to allow you to produce anything of real value or importance. Even if you break all the rules—blow up something, for example, or create an artificial life form—the stakes just aren't high enough. "So I blew up Bethesda," you might think, "it was just a school project."

If You're Angry with Someone You Can Threaten to Exterminate Him or Her

Pro: The layperson is very impressed by the voodoo of science. By casually referring to the vast resource of destruction at your disposal, you can persuade an object of your desire to notice you and perhaps even take an interest in you. If it's someone you dislike you can have him or her obliterated with the flip of a switch, or with a couple of cc's of something. Always remember: when you hold the future of the world in your hands you deserve respect.
Con: Deep-seated resentments by society at-large may make themselves felt eventually. Joe and Jane Average, after being held hostage for forty years, may get sick and tired of

science and scientists. And there's always the random sociopath with a chainsaw. You can threaten a sociopath with nuclear extermination, but that may have been a factor in the pathology in the first place. So you'd be back at square one, alone in the lab with a sociopath wielding a chainsaw. Is that what you want from a career?

Since People Mistrust Scientists in General, You Can Keep People from Attaching Themselves to You, Leeching Off You, and Wanting Things from You

Pro: The scientist who values himself and his time knows well the absolute necessity of keeping social contact to a minimum. People are a liability in a laboratory situation.
Con: It is in the best interests of humanity to keep science on an even emotional keel. This is a thankless and impossible task, but one to be striven for. Like science itself, we must all aim for progress and hope for perfection. After all, a scientist is human. When you prick him or her, does he or she not bleed?

□ *Why do scientists drink so much coffee? I just got my PhD, and I don't like coffee. Will I have to learn how to drink it now?*

■ Unfortunately, yes. Coffee is essential to any scientist, pots and pots of it. In order to function as a true scientist, you must possess what the layperson calls "coffee nerves." Science calls this *hypersynaptic calculosis*. What most people think of as the jitters is actually a state of creativity. The scientist who is not "jittery" is merely *thinking*. Thinking is fine as far as it goes, but it doesn't go far enough. You must make those great intuitive leaps, from the lowly atom to the mighty stars and back again, in split seconds. Coffee lets you do this. Of course your hand is usually shaking so uncontrollably you cannot read your own notes, but that's part of the price you pay when you become a scientist. If you can't pay that price, you'd better get out now.

□ *What does the acronym* FDIC *stand for?*

■ *FDIC* stands for Five Hundred Doctors in a Corner. Some are MDs, some are dentists, some have PhDs in economics. Besides guaranteeing loans, they are responsible for phrases like "Nine out of ten dentists recommend Trident for patients who chew gum." But what of the tenth dentist? The doctor who doesn't go along with his colleagues is driven from the corner in disgrace, to eke out a meager living endorsing diet plans and real estate scams or writing phrases like "Crest has been shown to be an effective decay-preventing dentifrice when used in a conscientiously applied program of oral hygiene and regular professional care." The doctor who wrote this committed suicide in 1956. I blame the FDIC.

□ *Where exactly is the scientific community? Can* anyone *live there?*

■ You must pass some very rigorous testing to be able to rent or buy in the scientific community. Located in the Pocono Mountains just outside of Port Jervis, New York, the scientific community is an ideal place to live, work, and raise a family. Unfortunately, most scientists have little interest in anything but work, so the delightful scenery afforded by the Delaware River goes unappreciated. And although the married scientist may have a family, chances are he has little contact with his children until they're old enough to use a slide rule. So I guess you'd have to say the scientific community is a quiet place, filled only with the weeping of neglected housewives and the occasional tantrums of a fatherless child.

□ *Why do scientists in the movies always have a foreign accent?*

■ This accent is usually German, and several theories have been advanced to explain this phenomenon. One theory has it that most German scientists are frustrated actors, and when jobs in the movies come up, they leap at them eagerly, even if it means they are forced to say things like "Zair are zum zings man vass not meant to know" or "Zoon, my darlingk, you vill be a bride off science." The other theory holds that there is something in the nature of film itself that causes a scientist to speak in a foreign accent. When the camera starts to roll, it causes a thin cold smile to play about his lips; it causes a cold gleam to come from his eyes; it may even cause him to cackle and rub his hands together in a sinister manner. This is only a theory, but I try to avoid being photographed if I can, just to be on the safe side.

NATIONAL ARCHIVES PHOTOGRAPH

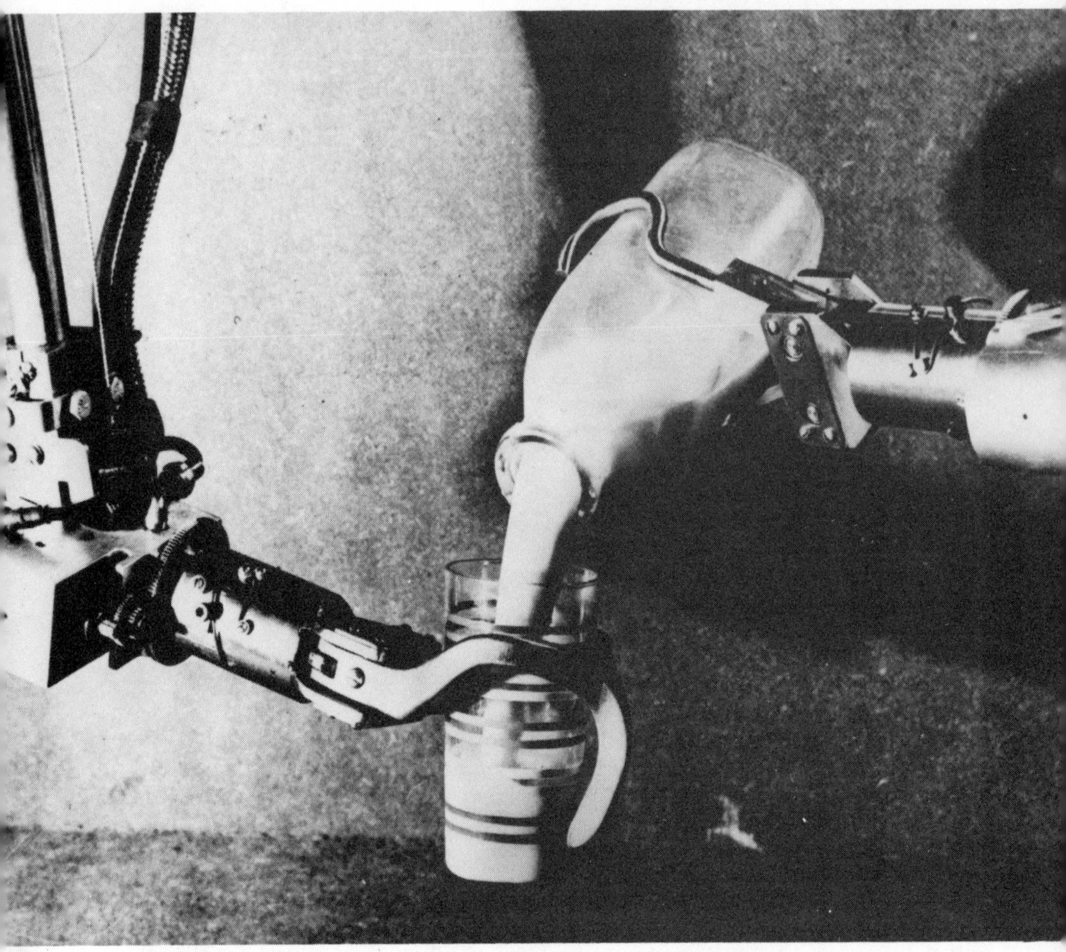

New Science makes the simplest actions complicated and fun!

Chapter Ten

How to Turn Your Home into a Laboratory

> Your home should reflect the very special person that is you. Now that you've changed, shouldn't your home change, too?

☐ *How old do you have to be before you're old enough to dissect a frog?*
☐ *How do you make a good hair-setting solution at home?*
☐ *Where does styling mousse come from?*
☐ *Is the scientific method a form of birth control like the rhythm method?*

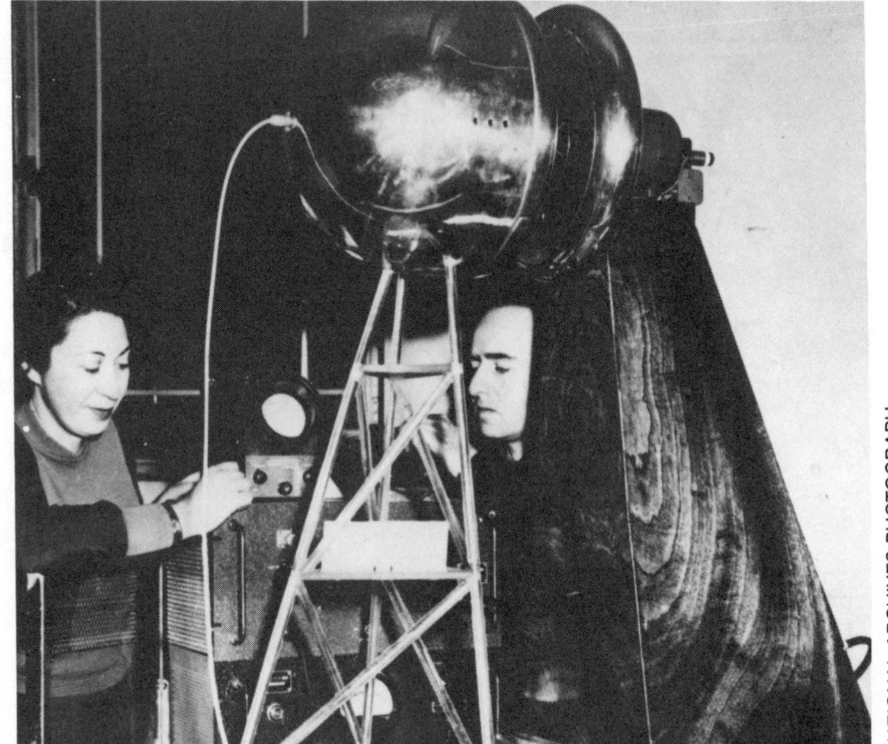

NATIONAL ARCHIVES PHOTOGRAPH

With a few simple adjustments your chifferobe can become a soothing sauna.

INTRODUCTION

It's easy, really. Any household gadget is a scientific instrument just waiting to be used. Did you know that your toilet is also a cyclotron? Your mixer a centrifuge? Even your stereo can, with a few simple conversions, become a radio telescope.

Few people have the insight and personal courage to make such a change. These are the same people who bitterly complain about the lack of laboratory space available to amateur scientists. If a copy of this book could be sent to each of these whiners and gripers, their problems would vanish overnight and my financial problems would vanish, too. Such is the power of one good idea.

TURN YOUR HOME INTO A LABORATORY

Step 1.

Remove your clothing. Stand naked in the middle of your living room. Put soothing music on your stereo (I recommend "Moonlight Memories" by the Jackie Gleason Orchestra). Survey your living room. Look closely at each piece of furniture. Think to yourself, "How does this resemble a piece of laboratory equipment?"

Sometimes the answer will come quickly. Sometimes it will be nothing more than an anxious pulling at your gut. But an answer will come, in every case, if you are patient. Now, put your clothes back on, and walk into your kitchen.

Step 2.

Cook yourself a big lunch. Force as much food into your mouth as you can. Now think of what just happened. A variety of food (and/or foodlike substance) left your refrigerator, were mixed on your stove, and ended up in your mouth.

The same thing happens in any working laboratory. The only difference between a laboratory and a kitchen is that in the lab the person who mixes the substances is not the person who ingests them.

Step 3.

Cover your windows with black plastic garbage bags. With this simple action you have turned your house or apartment into an isolation tank. This eliminates distractions and keeps prying eyes away from your experiments.

In such an environment you will do your best thinking. Spend a week or so here to digest what you ate in Step 2 and to think. This is called *brainstorming*. Your ideas are important. Somebody might want to know about them. And now you can use your common household furnishings to spread those ideas to the public at large.

The Home as Laboratory

The changes you have made in your home may not be visible to an outsider, but you have begun to see your home as a lab, and you are the person with whom we are concerned.

Sometimes the changes you've made will upset you. You're going to wish your home was a home again. When you wake at night and wander half asleep into the bathroom, the drowsy thought that you're entering a sterilized vacuum may upset you. This is normal. Science is frequently upsetting. And waking up in the middle of the night is often a terrifying experience, even to a nonscientist. You must take comfort in the fact that you are not alone.

Eventually, like me, you may give up sleeping altogether. This will give you twenty-four hours a day to work. And when you don't feel like working, you can scan lists of random numbers or listen to sine waves on the radio. All in the privacy of your own lab/home! That's what I call living.

So with the completion of three easy steps, you have laid the groundwork not only for a transformation of your home, but also for a transformation of your entire way of life.

Frightening, isn't it?

Take the time to be frightened. Truly, deeply frightened. When you emerge on the other side of this wall of fear, you will know what it means to be a scientist.

WHAT EVERY SCIENTIST SHOULD KNOW

Certain Phenomena

As you go through the process of becoming a scientist, certain phenomena will begin to occur. You will wake up one morning and notice a tenderness above your eyes. Your brow will seem unusually prominent. The top of your head will develop a certain pointiness. When this happens, relax. You're just turning into a scientist, that's all.

You see, there comes a time in every boy or girl's life when his or her true nature emerges. Like a slimy gosling emerging from a goose egg, the proto-scientist finds that nature has designs on his or her body. These designs seem shocking at first.

Unless you are aware of what's happening, you may think you are going insane! Don't worry. Many scientists *are* insane, but losing your mind is not a necessary stage in scientific development. The worrisome tenderness will pass. And your awkward smarty-pants attitude will evolve into a self-assured smugness. It just takes time.

But What Can I Do about This Discomfort Now?

The first thing to do is to stay calm. Getting upset will only make the tender areas more tender, and aggravate the scientific compulsion to invent something unnecessary or dangerous. So you must find some way to relax.

Unfortunately, the relaxation techniques used by middle-aged women in yoga classes are of little use to the young scientist in the first flush of post- or pre-pubescent graduative psychokinesis. No, for a fledgling scientist meditation just won't do. It smacks too much of spiritualism.

No, you must medicate yourself. I was helped through this very difficult time with chamomile tea. Steep about fifty grams of lightly irradiated chamomile blossoms in 500 ml of distilled water, and suddenly that unshakeable feeling that you're vibrating will go away. For a few minutes.

Over-the-counter drugs can approximate chamomile tea in their effect on the central nervous system. The fact that these drugs are expensive and come in colorful foil-lined packets is an assurance that is nothing to sneeze at either, especially if you're screaming into your pillow at night instead of sleeping.

SUGGESTED LABORATORY ETIQUETTE

The Laboratory is a tabernacle of investigative knowledge. No, make that a Tabernacle of Investigative Knowledge. As in any tabernacle, there is a code of behavior—call it an ethic, if you wish—that goes along with working and worshipping in such a Tabernacle. Science calls this code "Laboratory Etiquette." Like good table manners, Laboratory Etiquette says a lot about who you are and where you came from.

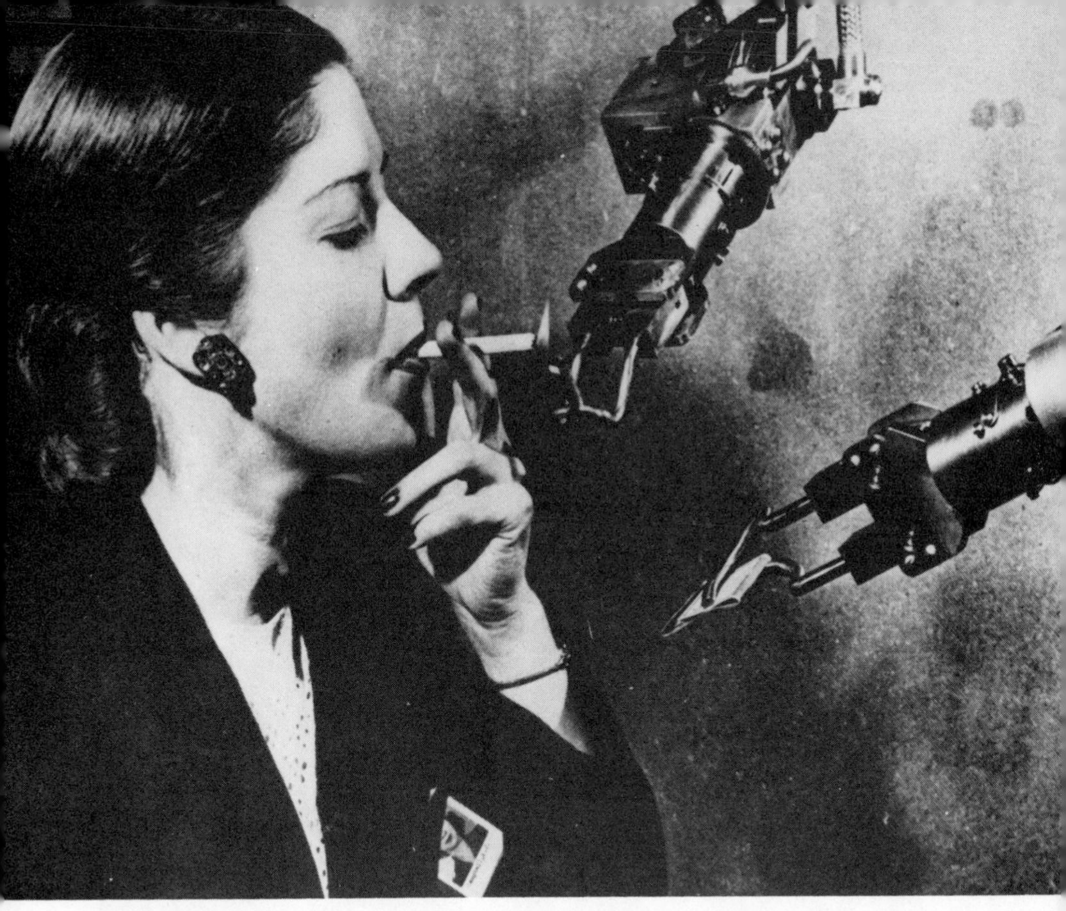

Even a robot should be a gentleman. Politeness counts.

NATIONAL ARCHIVES PHOTOGRAPH

Lab Dress

Proper attire consists of a spotless white smock, tied snugly around the waist. A tasteful and small insignia can be worn above the left breast. One's name is never embroidered on the smock; that is reserved for bowling shirts. If you must wear something that displays your name, a grey or blue formica nameplate may be worn. Under no conditions may this nameplate say, "Hello, my name is _______ ."

Grooming

Hair must be worn short. Hair creme and sprays are not allowable, both because they are possible contaminants, and because they project an image inconsistent with a man or

Strict segregation of men and women is necessary.

woman of science. The scalp should be examined daily for traces of dandruff. Shampoo twice daily with a mixture of shampoo and diluted bleach, then soak overnight in brine, then brush with an extremely fine-tooth comb. This is called nitpicking; it is not only hygienic, it is an extremely honorable and recreational scientific activity.

Social Intercourse in the Lab

Men and women should avoid working as laboratory partners. When together, the male and female minds cannot long concentrate on purely scientific endeavors.

What men and women do outside the lab is their business, provided they follow the grooming procedures described above. If communication between consenting scientists is necessary, it should be accomplished by written message or

TO DO AT HOME

A WORKING ATOM

It's easy to create a working atomic model in the convenience of your own kitchen. First, to duplicate the conditions of a science room, you must lower the temperature of your kitchen to absolute zero. To accomplish (or do) this, seal all entrances and windows, crank your fridge all the way up, then open your freezer and refrigerator doors. Be sure your hands and feet are warmly covered. Gloves are preferable to mittens, because you will be doing some close manual work.

In another room, using Easter egg dye and ordinary medium grade A eggs, you will have created your protons and neutrons. Dye the proton egg blue (for positive) and the neutron egg pink (for neutral). Rub the blue egg briskly *up and down* on a piece of wool to give it an electric charge, then attach it firmly to a coat hanger. Wrap this unit around the pink egg. Now dye another egg green to act as your electron. Rub this egg *sideways* on the piece of wool to create a negative electrical charge. Now place this green egg at the correct distance (it should be roughly a foot) from the nucleus of pink and blue eggs. The opposing charges should keep the green egg suspended in midair. Keep trying until you get it right. (*Note:* Make sure that these eggs are hard-boiled before you begin!)

Now, carefully juggling this model, take it into your kitchen. Cover your floor with shaved ice and a flour-and-water paste. This represents your quarks.

Now open the windows, turn off the refrigerator, and take pictures with an ordinary Polaroid camera.

This is how science spends a great deal of its time. Now you can, too, at a fraction of the cost!

CRT display. If speech is the only way a certain communication can be accomplished (a shout of "Fire," say, or "Eureka"), then it is permissible, if the scientist does not persist in this behavior for longer than is absolutely necessary.

SUGGESTED LABORATORY METHODS

A part of your laboratory etiquette may actually involve doing something *in* the laboratory. Remember, if it looks right, you're probably doing it right. Here are some more tips.

Measurement

Precision is of the utmost importance. Amounts weighed must be weighed on scales so sensitive that a fingerprint would be massive enough to affect a reading. Solutions must be micropippetted in micropippettes, which must be discarded after use, because they are very expensive and it's dramatic to throw them away after one use. It also creates jobs in the micropippette industry.

When you have carefully measured whatever you are measuring, add just a little bit more. This scientific version of poetic license is called "random error" and is very important to a branch of science called statistics.

Solutions should be stirred with a wooden spoon, or if none is available, a clean finger. Be sure to wash the finger afterward, or an absentminded suck might render the experimenter unconscious.

Handling Dangerous Substances

These are best handled at a distance. Hiring someone else to handle them is an attractive alternative. The preference is for a teenager desperate for a job, or someone who doesn't fully appreciate the danger of the work, an immigrant perhaps, or someone who doesn't speak English very well.

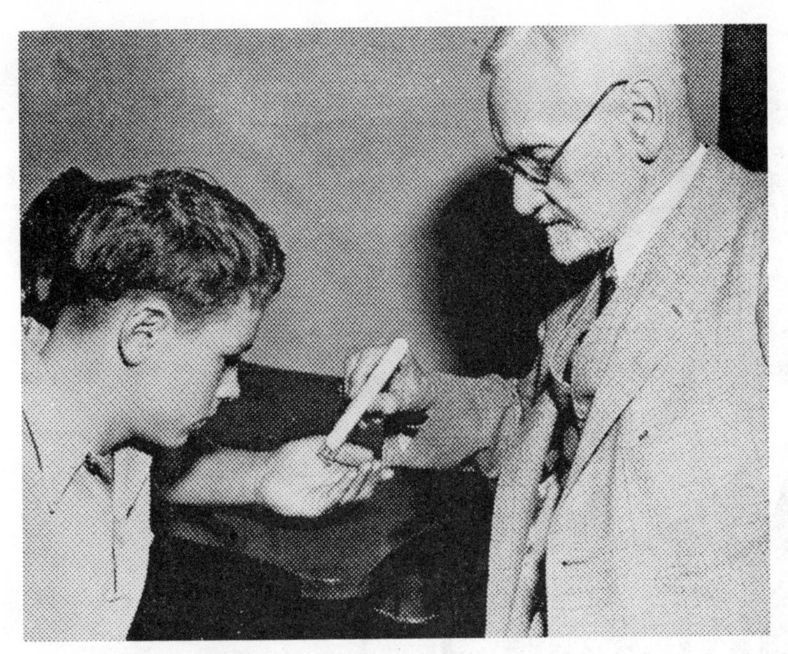

"This isn't going to hurt you."

Remember. You are the result of years of Yankee know-how, and the nation needs your knowledge so that future generations can benefit from the harvest you will reap with your mind. Let others less gifted provide the sweat of their brows. Let your brow be wrinkled in thought, furrowed with ideas which will later find their way into furrows plowed by those with less—never mind, you get the picture.

☐ *How old do you have to be before you're old enough to dissect a frog?*

■ Laws vary from state to state. In Oklahoma, frog dissection is not permitted until the age of forty. In Delaware, frog dissection is compulsory for grades K through 6, but in Wyoming it is not permissible at any age. I can't tell from your letter whether you're enthusiastic about frog dissection or dreading the experience. Nor do you say how old you are. I have to admit I've always enjoyed frog dissection. It's even sort of a hobby of mine, a pastime to while away the odd Saturday afternoon hour. This enjoyment probably stems from my high school lab partner, a girl named Melissa. She was my first date, and to this day, whenever I smell formaldehyde, I think of her.

☐ *How can I make a good hair-setting solution at home?*

■ A good hair-setting solution is chemically quite complex but fairly easy to make. Ask Mother for the following chemicals, which are obtainable in virtually every home. Just take a bottle of common laundry bluing. Strain it through a bolt of cheesecloth. Dad probably keeps it in the garage to polish the car. Then stir in 100 grams of trioxymethane, a compound often used to treat athlete's foot. Distill in a vacuum, then boil in a nitrous oxide atmosphere—easily obtained from the picture tube of the family color TV. By the next morning you should have a gummy, smelly substance that will make Mom's hair stand on end for a long time.

☐ *Where does styling mousse come from?*

■ Styling mousse is extracted from the Sassoon gland of the Missouri musk weasel. Cosmetologists would like us to think that it comes from a moose, or mooses, but the fact is

Missouri Musk Weasel (almost extinct).

that the gland of a moose produces styling gel, a different product altogether. Like its confectionery counterpart, chocolate mousse, styling mousse has no nutritional value. Misuse of styling mousse is a felony. So the next time you see a teenager with spiked hair, whisper a prayer of thanks that you don't have the glands of a dead weasel on your head.

☐ *Is the scientific method a form of birth control like the rhythm method, and if so, how does it work?*

■ Your ignorance appalls me. The scientific method is a way of gathering information to prove or disprove a theory. Let's take a couple of examples.

Let's say that men and women belong to the same species. It's easy to prove. A scientist takes a man and a woman, removes their clothing, places them in a room together, and watches them through a one-way mirror. After a few hours it should be obvious that men and women are indeed members of the same species. And another scientific law goes into effect.

Let's take another thesis: Dogs and angleworms are the same creature. This silly theory is easily disproven. Put a dog and an angleworm in the same room and observe their behavior. The dog will bark, whine, and sniff the worm, and the worm will squirm along the floor looking for dirt.

Scientists do this sort of thing all the time. They don't have time for sex.

WHAT WE'VE LEARNED SO FAR

The world is full of wonders. You can have a piece of those wonders by becoming a scientist. Science is a painful process with many rewards. Anybody can be a scientist, with proper grooming.

REVIEW QUIZ

1. Most New Scientists wear beards. If you are a woman who desires to become a New Scientist, must you grow a beard as well?
2. You experience a certain nagging feeling. Your head feels like it may explode. When you lie down you feel nauseous. These symptoms are peculiar to (a) werewolves, (b) New Scientists, (c) lobbyists for the NRA, (d) all of the above.
3. Are soybeans an effective meat substitute?
4. What do these handy phrases have in common?
   A. "Soon, soon, my darling, you will be a bride of science!"
   B. "They laughed at my theories at the Academy, but tonight they will laugh from the other side of their faces!"
   C. "So you see, Johnny, when the kitten is attached to the pendulum, the pendulum remains motionless."

WHAT WE'VE LEARNED SO FAR

D. "The Crab Nebula, though visible to the naked eye, is actually billions and billions of miles away."

5. Is a cyclotron an effective means of making yourself a tall cold milk shake?

Part Four

The Final Exam

NATIONAL ARCHIVES PHOTOGRAPH

NATIONAL ARCHIVES PHOTOGRAPH

Chapter Eleven

New Science Pays Off

IN THE MOOD?

You are standing on the threshold of a bold new career. There's only one more step to take, and that of course is the final examination. So sharpen your pencils, put on your thinking cap, and take your seat. Answer each question firmly and correctly. Don't peek at the answers (*on pages 172–78)* until you're done.

We're on the honor system around here, but it may help to imagine my stern presence before you, pacing back and forth in front of a cracked and dusty blackboard. With one hand I am clasping my left shirt pocket, which holds a pocket protector, which holds a wide variety of ballpoint pens, none of which work. The shirt is white nylon, and you can see the straps of my undershirt through the material. My pacing distracts you from the test. I am wearing brown shoes.

To get you even more in the mood, imagine the wooden desk in front of you, scarred and massive, holding more shellac than wood and bearing the initials of hundreds of

long-gone students, each carved initial containing a stiff pool of dusty shellac. You may, if you wish, imagine a more intelligent student across the aisle from you and sneak a peek now and then at what she is writing.

All right. In the mood? Let's begin.

FINAL EXAM

Part I: Multiple Choice

1. We put air in basketballs to make: (a) babies, (b) antigravity, (c) cothurni, (d) ancient Greeks.
2. When we burn a stick, we are: (a) having fun, (b) making phlogiston, (c) falling in the forest with nobody there to hear us, (d) ignoring forest safety.
3. When electrical appliances wear out, we should: (a) use them anyway, (b) thank the higher powers, (c) wear cothurni, (d) sit in the dark.
4. Waste materials piled in out-of-the-way places may: (a) move around at night when no one's looking, (b) implode, (c) be stolen, (d) be useful.
5. Insects are: (a) disgusting, (b), useless, (c) many-legged, (d) all of the above.

Part II: Fill in the Blank

________ is formed when we rub two ________ together. ________ can live for days without ________. We call a ________ with ________ a ________. ________ is a simple ________ that cannot be ________ any smaller. ________ was the father of ________. The number of ________ that can dance on the head of a ________ is ________. There are ________ laws of ________.

Part III: Mix and Match

Below are ten sentences with the subjects and predicates mixed up. Match each subject with the correct predicate to make sentences in your notebook.

SUBJECTS

1. Boiling water for thirty minutes
2. New Science
3. Old Science
4. Green scum on water
5. Ancient Greeks
6. The scientific method
7. Coffee
8. Electromagnetic radiation
9. Natural gas
10. The theory of relativity

PREDICATES

a. is often piped thousands of miles for home use.
b. is our most valuable fuel.
c. causes many explosions at home.
d. is a myth.
e. is our major source of light and heat.
f. didn't know much.
g. is a constant source of inspiration.
h. is a constant source of inspiration.
i. is an ancient form of energy.
j. were stupid.

Part IV

Answer the following questions as best you can, without looking at the answers.

COLUMN A	COLUMN B
1. How many reasons are there to go on living?	1. Four.
2. If you put a cat and a dog in a cyclotron, which one would die first?	2. Dog.

3. What major contribution did electrical engineering make to the cosmetics industry just after the Korean War?	3. Hair creme, which had been used to lubricate the Norton bombsight.
4. If you were trapped on a desert island with a scientist, which one of you would crack first?	4. You.
5. If you were trapped in a science room, would your constitutional rights be violated?	5. Yes, but it's all in the interests of science.
6. What is the best way to defend yourself against an arrogance attack?	6. Respond with arrogance.
7. Why are so few scientists also ordained ministers?	7. They don't have the patience for Bible study.
8. If a scientist had the choice between splitting an atom and eating a good meal, which would he or she prefer?	8. The atom.
9. Why?	9. More fun.
10. Would it depend on the meal?	10. No.
11. The type of atom?	11. No.
12. Are the laws of gravity different in a Popeye cartoon than in a Bugs Bunny cartoon?	12. They are all the same.
13. If you were dropped out of an airplane above central Alaska and had on only a T-shirt, Bermuda shorts,	13. None.

and flip-flops (ZORIS or COTHURNI) for shoes, what scientific laws could you use to help you survive until you could be rescued?

14. Does whatever makes your blood red also cause the salmon to be red, or is it something else entirely?

    14. Something else.

15. Is there something unsavory going on inside the new automatic-focus cameras that would cause the average photographer to think twice before using such a camera if he were only aware of this unsavory something's existence?

    15. No.

16. Which weighs more, a kilogram of lead or a thousand grams of antimony?

    16. This is a trick question. They actually both weigh *less* than a ton of feathers.

Part V: Essay Question

Prove the theory of evolution. Take as much time as you wish. We'll see you back here in a couple of weeks. Don't peek at the proof.

Part VI: Arrogance Potential

We have come to the final section of the examination to determine your worthiness for a coveted master's degree. We will forego the oral examination; my lack of social skills makes it difficult for me to indulge in idle patter. We will forego the thesis. I wouldn't bother to read it, frankly. Instead we will do this.

"You know my methods," as Holmes said to Watson. Take the following questions and answer them as I would. If you do indeed answer them as I did, you are worthy of that awesome degree that I myself possess.

Begin now.

1. Is there any connection between the Krebs Cycle and a woman's menstrual cycle?
2. When you make bread, does yeast feel pain?
3. Is a molecular bond the same thing as a tax-free municipal bond?
4. Which is correct: 8 + 5 *is* 12 or 8 + 5 *are* 12?
5. Can you peel potatoes with an all-purpose flashlight?
6. What's the difference between a Moebius strip and the Las Vegas strip?
7. Why do people twitch as they are falling asleep?
8. Why do we have Celsius and Fahrenheit?
9. What is the difference among germs, bacteria, and viruses?
10. Can one name a new element after one's dog (Fluffy)? How?
11. What happens to water, exactly, when it freezes?
12. Many people realize they can tell the temperature by counting the chirps a cricket makes. But how does the cricket know what temperature it is?

FINAL EXAM ANSWERS

Part I

1. (b)
2. (d)
3. (b)
4. (d)
5. (d)

Part II

Phlogiston, ancient Greeks. Lavoisier, Maxwell's Demon. Bilogist, a master's degree, wistful scientist. Electricity, evil, made. Washington, our country. Angels, pin, thirteen. Five, traffic safety.

Part III

1. (a) Boiling water for thirty minutes is often piped thousands of miles for home use.
2. (g) New Science is a constant source of inspiration.
3. (c) Old Science causes many explosions at home.
4. (h) Green scum on water is a constant source of inspiration.
5. (f) Ancient Greeks didn't know much.
6. (i) The scientific method is an ancient form of energy.
7. (b) Coffee is our most valuable fuel.
8. (d) Electromagnetic radiation is a myth.
9. (e) Natural gas is our major source of light and heat.

10. (j) The theory of relativity were stupid.

Part IV

The answers are right next to the questions. You can peek now.

Part V: Proof of the Theory of Evolution

The theory of evolution is easily proven. You can actually create life in the privacy of your own home. Just follow these simple steps:

1. Take a shower, but don't clean the clogged drain. Mixed in with that damp tangled hair is what the layperson knows as dandruff, but what are actually DNA molecules. Certain generic brands of hair conditioner can trigger cell division. Experimentation will show you which conditioner is right for you. Styling mousse, however, will halt the process.
2. After a week or so, if you've found the right conditioner, you should have a small green thing in your tub. It should have the appearance of algae and the consistency of green Jell-O. Let it sit.
3. After a month you should have a throbbing green throw rug. If you're very patient (and don't mind bathing with an artificial life form), you should end up with a seven-foot-tall green hulk. While this creature is none too bright, it will be totally devoted to you and will even obey simple commands such as "Sit," "Roll over," and even "Kill."

Try it! That's what science is all about!

Part VI: The Right Answers

1. Is there any connection between the Krebs Cycle and a woman's menstrual cycle?

The Krebs Cycle is a biochemical model that explains how plants turn carbon dioxide into oxygen and water in the presence of sunlight. Menstruals, as we all know, were performers in blackface who played the banjo and slapped tambourines. Like their medieval predecessors, these menstruals rode bicycles with the top bar missing, the better to carry their banjos and watermelons.

There were no women menstruals, at least none who could ride a bicycle, and the Krebs cycle was discovered by a man, Maynard G. Krebs, who later went on to discover the so-called Gilligan Effect. But that's another story.

2. When you make bread, does yeast feel pain?

Definitely. In fact, the tiny bubbles in bread are yeast screams, which are released when you cut the bread. You can't hear them because they're at too high a frequency, but your pets can. This is why so many dogs seem to be begging at the dinner table. They aren't begging for food; they're begging for the life of yeast.

In a recent study, yeast was given a growth hormone by the Pillsbury people. This yeast, known in layman's terms as "POPPIN FRESH," emitted a scream when a lab assistant poked him in the stomach with a forefinger. Pillsbury used this demonstration in commercials but put happy music behind it and cut away before POPPIN FRESH passed out from pain. All part of their program to convince us that yeast enjoys getting baked alive. If yeast pain upsets you, I recommend you do what the Buddhists do. Eat rice. Or do what I do, and eat only things made in a laboratory.

3. Is a molecular bond anything like tax-free municipal bonds?

Yes, they are exactly the same. If you wish to create a chemical bond, you must approach the various elements by following the laws of nature. And the laws of nature are very much like a planning commission. You approach the city commission with a bond to build, say, an airport. You approach the laws of nature with a bond to build, say, hexachlorophene. In both cases the people who ultimately pay for the project are yet unborn. But once you get past that first commission you've got it made.

4. Which is correct: 8 + 5 *is* 12 or 8 + 5 *are* 12?

Neither. The correct answer is "8 + 5 *equals* 12," but only if the system of addition used is what I call the "There's something wrong with this picture" school of mathematics. This new math was developed by the people who make those fun puzzle brainteasers you see on place mats in truck stops. With this new math, you can feel free to say "2 + 2 is kind of 4," or even "5 × 5 is like 26, more or less." Try it. It'll make balancing your checkbook easy and fun.

5. Can you peel potatoes with an all-purpose flashlight?

Consumer fraud is a bit out of my line. All I can do is sympathize. I've faced this problem myself. The pants suit that said, "One size fits all," but ripped at the seams after only four people. A blatant rip-off. The paper towel dispenser that said, "Rip down, tear up," and when I ripped it down and tore it up, my hands were still wet, and there were reams of low-grade paper and mangled machinery all over the floor. And what about those cans that say, "Contents under pressure"? Why should I buy a can with anxious contents? There's enough stress in the world, and when I buy a can it better have calm contents or nothing at all. I recommend you take your flashlight back to Hong Kong and try to get your money back. Good luck.

6. What's the difference between a Moebius strip and the Las Vegas strip?

None, really. A Moebius strip is a strip that has been joined together in such a way that its two surfaces become one continuous surface. There is no in and out. It's also difficult to discover the ins and outs of the Las Vegas strip, and if you've ever been to Vegas you know that it, too, is all surface. The main difference, if you can call it that, is that strippers can strip in Las Vegas but not on a Moebius strip. The Moebius strip can be used as a no-pest strip, but pests in Las Vegas are impossible to avoid. Gambling is legal on the Vegas strip. I suppose you could gamble in Moebius as well, but a playing card would be the same in front as in back, so it wouldn't be much fun.

7. Why do people twitch as they are falling asleep?

As one falls asleep, one is jolted by the high-voltage shocks emitted by the hypothalmus. This organ, a neighbor of the pituitary gland, is a more evolved cousin of the electric eel. As we fall asleep we de-evolve through millions of years of genetic memory, ending up as pure protoplasm. The wet spot one finds on the pillow, especially after a nap, is nothing more than the residue of simple amino acids leaked by our proto-selves during sleep. Elementary gas chromatography will bear this out. So the next time you wet your pillow, take the pillowcase to the nearest lab and have them check it out. And tell them Dr. Science sent you.

8. Why do we have Celsius and Fahrenheit?

It's Ms. Celsius, actually—Linda Celsius and Bob Fahrenheit. They were a couple in Massachusetts seeking a divorce. As you know, scientists have named everything from cars to rockets to planets after ancient and obsolete deities. By the time accurate temperature recording became a reality, these names had all been used up. That's when meteorologists turned to Bob and Linda, who set aside their bitter differences long enough to put their names on your thermometers.

9. What is the difference among germs, bacteria, and viruses?

Germs are invisible, disgusting creatures that live on toilet seats and dirty hands. If they can find their way into your mouth, they can kill you. Bacteria, on the other hand, are elegant, educated city cousins of germs and have more subtle ways of infecting you. They can enter your body through any orifice. Vira (the plural of virus), are mutated quarks sent by the Powers of Darkness to create a sort of hell on earth. Soap can kill germs, antibiotics can kill bacteria, but nothing can kill a virus. Nothing short of exorcism, that is, and that in itself is often fatal. So good luck.

10. Can one name a new element after one's dog? How?

First you have to have a trademark search done. This is to ensure that the name ("Fluffy," say) is not already a registered trademark. If it is, then you must either rename your dog or sue the offending party in a court of law.

Then you must approach the International Tribunal of Element Names, which meets every fourth Shrove Tuesday in Zurich. It doesn't matter, incidentally, who discovers this element—you, Fluffy, or some overworked scientist. What matters is who can show the tribunal members the best time in Zurich. That's hard to do in the most boring city in Europe.

Good luck, and I'll be looking for "Fluffy" on my periodic table. That's the kind of guy I am.

11. What happens to water, exactly, when it freezes?

Freezing is a myth. What we call *freezing* is actually a by-product of natural evolution. It's a form of protective coloration. Water assumes the character of ice whenever danger threatens. This danger usually takes the form of cold weather. Ice and water are actually quite different things. Water is everywhere; it's free and flows easily. True ice is harder to find but can usually be found in small buckets in hotel rooms. You can sometimes find ice lurking in large blue plastic bags in the frozen foods sections of your grocery store. You can capture ice if you run screaming into the frozen food section, leap, and tackle a blue bag before it can get away. Next time you're in the store, try it. Tell them Dr. Science sent you.

12. Many people realize they can tell the temperature by counting the chirps a cricket makes. But how does the cricket know what temperature it is?

While you're on the verandah swatting mosquitoes and complaining to your friends about how hot it is, the cricket is sitting in air-conditioned comfort watching the six o'clock news. Out of boredom, perhaps, or a genuine need to give us information, the cricket communicates this weather information to you. The cricket will also click out, in Morse code, the final sports scores and national headlines, even such phrases as "Now this," "Film at eleven," and "Our White House correspondent filed this report." Some scientists call the cricket "the Walter Cronkite of the insect world," which is accurate but somewhat silly. After all, you've never seen Walter Cronkite rubbing his legs together. At least I hope you haven't.

Awarding the Degree

This is a solemn moment. Don't spoil it by whispering and giggling. For the rest of this page stand at attention, your eyes forward, hands clasped firmly behind your back. Let the presence of the throng of students around you fill your heart with pride. Thousands of flashbulbs ignite in your peripheral vision as proud Moms and Dads snap photographs to tuck away in their scrapbooks. Bored brothers and sisters fall asleep or sneak away in droves. Below you on the grandstand hundreds of distinguished professors, robed in black, stand rigidly at attention as the orchestra plays the stately "Pomp and Circumstance" over and over and over again.

This is it. You may have cheated. You may have stabbed your fellow students in the back to get here. None of that matters now. Just wait until your name is called over the loudspeakers, then walk with a determined stride down the center aisle. Receive your degree with a modest half-smile and a murmur of gratitude, then lightly shake the hand of the presenter (Lorne Green, if we can get him), and return swiftly to your seat. Do *not* resume your seat until the ceremony is over. The ceremony seldom lasts more than five hours. You can do it. You're a New Scientist now. Take your degree, and let's go to work.

Masters of Science Degree

____

has been awarded the Masters of Science Degree by a university that wishes to remain anonymous.

Rodney

Dr. Science

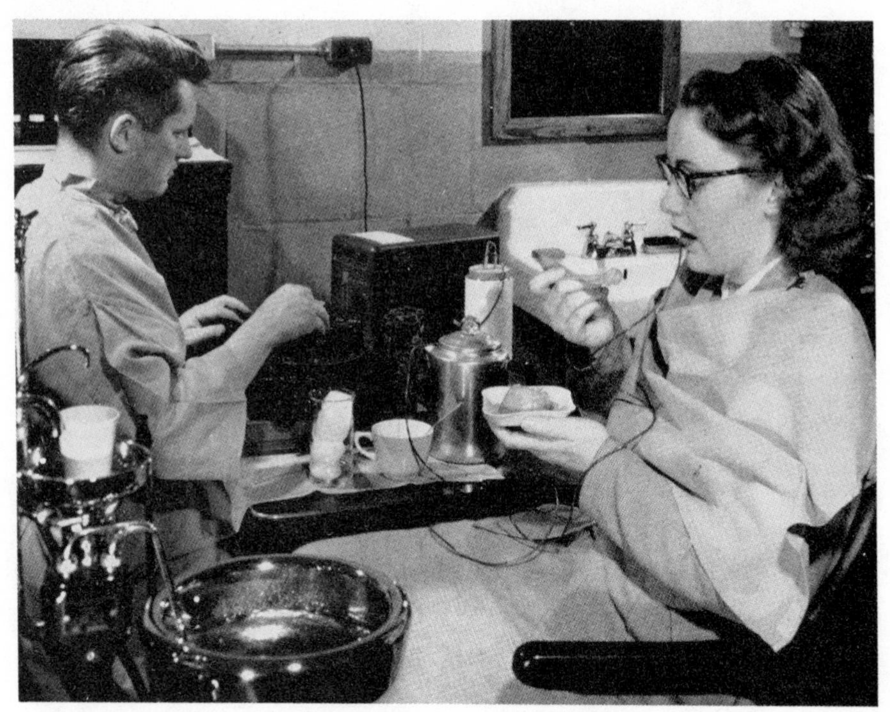

A husband and wife research team try to find her pain threshold. He notes the data while she gets ready to hurt herself.

Author's Epilogue

Congratulations! You've just finished reading what will probably prove to be the most important book you've ever read. I'd go so far as to say it will prove to be the most important book you'll ever read, but that would suppose that I won't write another. I don't suppose that, and neither should you.

There will always be new questions to be answered, new myths to be debunked. The minds of most Americans are like adjoining acreage in a great unplowed field. If this field could be plowed, seeded, and properly fertilized, the harvest could drive food prices down for good. Sure, the farmers would suffer, but there are more of us than them, and we like low prices in our *lifestyles*.

When we were children, we were promised a world made carefree, efficient, well-lighted, warm, sanitized, and odorless. Science promised to give us this world. But there seemed to be a hidden cost for all this progress. We had to give something to science. We had to sell our souls to science in order to realize the promises science made to us.

Or so we thought. Now we know that we had misjudged science all along. Science is happy just to be there, like a chubby friend, in case we need him. Science won't speak until spoken to. Science will never command, only suggest. Science doesn't need to make outlandish promises to earn our friendship. Science is already our friend. Always was, always will be.

And like the gentle inward voice that lets us know when we're on the right track, science is always reminding us to listen to authority, to believe experts, and to pay attention. You've read this book. You've done your job. Now leave me alone so I can do mine.

—Dr. Science, 1986
at the Fortress of Arrogance

Appendix A

The Importance of Ridicule

It's a scientific rule of thumb that every great idea must be ridiculed. The process is necessary, not only to build character, but to add a little spice to a scientist's dull life. When dull Charles Darwin first uttered his theory of evolution, he was hounded for weeks by gorilla-suit clad hooligans, who laughed raucously and called him "monkey man." Einstein's famous formula "E = MC2" was also greeted with hoots by a gang of rowdies, also clad in gorilla suits. Why they wore gorilla suits is not clear; perhaps rowdies get a discount on them. We don't know. We only know that the importance of mockery in science cannot be overemphasized. Behind every great scientist stands a joker—the obnoxious fellow ready to stick the "Kick Me" sign on Galileo's backside, the punk who puts two fingers behind Ben Franklin's head, just as the photographer snaps the picture.

Today, of course, science is more respectable. We don't have great scientists any more, just great research teams. It's hard to mock a committee. Still, we must remember that catcalls, Bronx cheers, and hurled rotten tomatoes are all part of the scientific method. If we want great scientists, we must also have great sarcasm. An idea just isn't an idea until it's been attacked.

You can help. Many scientist's names can be found in your yellow pages; check under Taxonomy or Genetics for starters. Call a Taxonomist on the phone and when he or she answers, shout "You have your head screwed on backwards," and hang up. Or write a scientist a long letter in crabbed spidery handwriting, tearing a precious theory to pieces. Scientists will be annoyed at first, then exhilarated. And if you can show up at a scientist's doorstep in a gorilla suit—well, your place in the history of Science and dialectic will be assured.

Appendix B

Father of "Father of..."

If you have any memory of science history, you will remember that every famous scientist is also the "Father of" something. Einstein is the Father of Relativity, for example; Darwin is the Father of Evolution. Some scientists did not want this title. Pasteur, when dubbed "Father of Sterile Hospital Techniques," tried to move to another city. Unfortunately, his paternity was proven unequivocally in a court of law, and he had to rear Sterile Hospital Techniques until it was old enough to make it on its own.

This whole fatherhood business may be one of the reasons why there so few women in science. After all, even with the liberation movement, women cannot be fathers. And to say, "Mother of Post-Modern Elemental Conversion Methods" just doesn't sound right.

It was Presocrates who first coined this term. He gave his followers new titles every day, such as "Father of Small Pebbles," "Father of Muggy Day," "Father of Light Breeze Coming from the Northwest at Five Miles per Hour." At first thrilled, his disciples became rapidly bored with his incessant title-making. As they walked away, Presocrates could be heard shouting, "Father of Walking away from Presocrates at a Right Angle to the Road!" "Father of Walking away from Presocrates with Angry Expression on Face," and so forth, until they were beyond hearing.

Today, Presocrates is not known as the Father of Anything, except in this brief entry. When you have read this I advise you to forget it. Make room in your grey cells for more important facts.

Appendix C

Old Wives

Many important scientific discoveries have been linked to what we call "Old Wives' Tales." Such universal truths as, "nature abhors a vacuum," or "lightning never strikes twice in the same place," have all been examined with care by the Old Wives Research Team, located in an underground bunker in Bethesda, Maryland.

Through trial and error, these painstaking researchers have examined and catalogued every aphorism ever uttered. Thus we have learned that nature actually rather likes a vacuum; old wives are the ones who abhor a vacuum. Vacuums are filthy, you see, and noise is caused by dirt. The incessant noise of *vacuua* kept the old wives awake at night; thus, these insomniac women began to tell tales to pass their sleepless hours. Today, modern conveniences such as the vacuum cleaner and space shuttle keep *vacuua* clean as a whistle and old wives are getting all the rest they need.

So now that the old wives are catching up on their sleep, we have learned that a watched pot does in fact boil, the good don't die young, and that not only can you put a cart before a horse, you can lead that same horse to water and force that horse to drink. These are scientific facts. Old wives snooze; science never sleeps.

We have also learned how much exactly is saved with a stitch in time (closer to ten than nine), and that you can cross the same river twice. The same river, by the way, is much safer than the same place in a thunderstorm, but you can't cross the same place twice, unless you have a bird in the hand, a bull by the horns, or a gallon of elbow grease.

Appendix D

Subatomic Particles

The twentieth century discovery of tiny subatomic particles heralds the end of the physical world as we know it. To the layperson this news might seem depressing, but the physicist has no such attachment to his or her discoveries. Physicists just smash atoms and let the quarks fall where they may. To those of you who haven't the time or special equipment to check out the atoms around you, here's a brief overview of where the atom stands, particle-wise:

To avoid confusion remember that a *lepton* is not a type of tea and a *boson* is not a buffalo. Many physicists have confused the boson with the bison, with disastrous results. It's actually quite easy to tell the difference. It's difficult to get a buffalo into a linear accelerator for one thing, and even if you could, it's very hard to clean the equipment after the experiment. Just keep in mind, that if the object in question does *not* have horns or mange, it is probably a boson.

And what do these subatomic particles do? Well, consider the atom as an offshore corporation. The electrons are the secretaries, busy answering the phones and answering correspondence. The neutron is the Chief Executive Officer, making the important decisions that keep the atom in the black. Neutrinos come and go, much like a messenger service, while the quarks and other subatomic particles work way down in the basement.

What quarks do down there is anybody's guess. Quarks are theoretical bits of matter, only there to make quantum theories taste good. The quark is visible only by the tracks of its decay. In other words, if you can't see it, it's there.

Appendix E

Glossary

Abdomen: That part of the human body that causes the most trouble in middle age; the most disgusting section of an insect.
Atom: Basic unit of life; bomb ingredient.
Absolute Zero: About as cold as you can get.
Absorption: The passing of digested food into the bloodstream; the mental attitude of a scientist performing an important experiment; what sponges do.
Acid: Any substance that turns litmus paper red.
Air: What we breathe.
Ampere: Mythical unit of electrical measurement.
Alcohol: Misunderstood vitamin.
Aneroid Barometer: Outmoded piece of equipment.
Alloy: French for "let's go."
Accelerator: Makes cars and neutrons "go."
Algebra: Outmoded form of mathematics.
Ammonia: Stinky yet excellent cleaning fluid.
Anomaly: Rare tropical flower.
Antibiotics: Sturdy and brave infection fighters.
Antidote: A brief, amusing story.
Avogadro Number: One fifty a pound, in season.

Azimuth: A rare tropical flower.
Bacterium: Tiny instigator of disease.
Baking Powder: Substance of obscure origin and unknown function.
Baking Soda: Primitive rocket fuel; refrigerator freshener.
Ballistic Missile: Scientist's joy.
Barium: Substance traced in barium traces.
Bath Salts: Delicious shower spice.
Battery: Electric prison; a felony.
Bauxite: Source of aluminum.
Benzene Ring: What scientists exchange when they marry.
Bessemer Process: What scientists complete on their honeymoon.
Biochemistry: One of the wistful sciences.
Black Body Radiation: Planetary orgasm.
Blood: Thicker than water.
Boiling Point: What many scientists reach when findings are questioned.
Brownian Movement: A saunter distinctive to many pedestrian scientists.
Bunsen Burner: The only safe method of Bunsen disposal.
Byte: Unit of information.
Caffeine: Source of inspiration.
Calcium: Mineral found in deposits, usually located in elbows or Wyoming.
Calculus: Outmoded form of mathematics.
Calorie: What is burned when cells ignite.
Camera: Catalyst for vacation fun.
Carbon Dioxide: Secret ingredient of soft drinks, champagne, and beer.
Cathode: Roman Catholic child before baptism.
Cell: Unit of life.
Cellulose: Plastic seed.
Centrifigal Force: Energy used when the face grimaces.
Chemistry: The study of chemicals.
Chlorine: Pool ingredient.
Chromosomes: Chrome in its larval stages.

Circle: Type of logic.
Cloud Chamber: Where internal combustion originates.
Colloid: Gelatin.
Comet: Smudge in the sky, with a tail.
Compass: Handy direction finder.
Computer: Data factory; smarter than humans; solver of all known problems; balancer of checkbook; word processor; efficient memory; greatest boon mankind has ever known.
Copper: Element; policeman.
Cosmic Rays: One of Lex Luthor's many secret weapons.
Critical Mass: What the earth would reach if God were to try to lift a rock so heavy He couldn't lift it.
Curie: Doctor's response to disease.
Cyclotron: A gobot who can turn himself into a bicycle.
Decibel: Unit of loudness—ten decibels make a whisper, forty make a shout, etc.
Degaussing: Famous French physicist.
Degree: Result of college education.
Denature: To castrate alcohol.
Density: Mental quality.
Depth of Field: Trajectory followed by a football.
Detergent: Wonderful substance that gets clothing crisp, white, and clean.
Deviation, angle of: A formula that when mentioned, causes much snickering in classrooms.
Diffusion: How particles travel.
Digital: Replaces old-fashioned analog.
Discharge: To fire.
Dispersion: Another way particles can travel, if they have the money.
Doppler Effect: The annoying whine of a passing train.
Ductless Gland: Useless organ.
Dynamo: A scientist who uses caffeine.
Echo: Type of sound.
Eclipse: What New Science will soon do to Old Science.
Electricity: General term used to describe the greatest evil the world has ever known.
Electromagnetic Radiation: The force that keeps tiny pieces of fruit attached to refrigerators.

Elementary Particles: Basic units of matter. *See* Appendix.
Elvis: God.
Emission Spectrum: A gasoline rainbow.
Energy: Matter after boiling point.
Entropy: Universal boredom.
Expansion of the Universe: What happens when the universe is inflated.
Factor: Ingredient.
Fallout: A nasty side effect of nuclear energy.
Fatty: Vicious nickname.
Feedback: Outmoded form of musical expression.
Ferric: Ironic.
Field: Where cows graze, unless they are outside gravity.
Fission: What atoms do for fun.
Formula: Scientific shorthand.
Fossil: A myth.
Fuller's Earth: Type of soil invented by R. Buckminster Fuller.
Fusion: What atoms do when they're bored with fission.
Gallon: Spanish sailing vessel.
Gamete: Baby cell.
Gamma Rays: How grandmothers travel through space.
Gas: Fun.
Gene: Famous cowboy actor.
Genetic Code: Another name for the so-called "Code of the West,"; e.g., "He who lives by the gun dies by the gun." his code was developed by Gene in 1947.
Geology: Study of the earth's crust. Why anybody would want to be a geologist is beyond me.
Globulins: Protein trolls that haunt laboratories.
Glue: Force that binds all matter together.
Gold: That glittering substance for which so many have died.
Greek Fire: Phlogiston.
Grid: Humorous-sounding word.
Gyroscope: Oven used to make Greek food.
Heat: That which, if not taken, forces one to leave the kitchen.

Hemoglobin: Mischievous imp who haunts laboratories.
Hertz: Unit of frequency; car rental company; prime ingredient in donuts.
High-Fidelity: Primitive CD.
Higher Power: *See* Elvis.
Hologram: The ghosts of cartoon animals that haunt amusement parks.
Hormones: Alarming secretions.
Humus: Type of dirt.
Hydrogen: Bomb ingredient; high school classes make it bark.
Ice: Summer tea ingredient.
Ignition: Patron saint of automobiles.
Implosion: What will happen to you if you leave the rocket without your suit.
Impulse: Motivation to shop.
Indigo: Mood.
Inertia: Basic state.
Ion: Highly charged particles that leak from unplugged sockets. Very dangerous.
Irreversible: Type of vest.
Isobar: A place where singles gather.
Jasper: What Gene's sidekick calls another cowboy.
Joule: Semiprecious stone.
Kilogram: What scientists send instead of postcards.
Kinetic Energy: The effect of too much caffeine in the system.
Lanolin: Marvelous ointment.
Laser: Weapon used on George Lucas films.
Latex: Type of sap.
Lecithin: Fat lite.
Lens: Plural of *len*.
Linear: Old Science way of thinking.
Liquid: Wishy-washy form of matter that can't make up its mind whether to be a solid or a gas.
Lysol: Stinky yet effective disinfectant.
Machine: Prime ingredient in the making of submachine guns.

Magnification: Size rays produced by microscopes, which make tiny objects larger.
Malonyl Urea: Rare tropical plant.
Maser: A weapon that never made it in the movies.
Mass: An abbreviated state.
Matrix: Ingredient used by computers to compose rough drafts of letters.
Matter: Frozen energy.
Mean: What one must be sometimes if one is to reveal the truth.
Microphone: Scientific instrument used to explore the nature of sound.
Mirrors: How magicians do it.
Modulation: How the Golden Rule says we should do all things.
Molecule: A bunch of atoms all glued together.
Mutation: Biologist's hobby.
Nadir: The pits.
Neanderthal: Critic of Dr. Science.
Neuron: What we will need when Ron retires.
Nickel: Doesn't buy much anymore.
Normal: Theoretical state.
Occlusion: Antonym of conclusion.
Octane: German for "eighteen."
Ohm: What one chants to obtain Nirvana.
Opacity: Property of fat glass.
Optics: Tiny insects that feed on sunglasses.
Organic Chemistry: Chemistry using compost instead of artificial fertilizers.
Osmosis: How Dorothy got back to Kansas.
Oxygen: Divers carry this in special tanks to give them life as they swim beneath the seas.
Ozone: Oxygen lite.
Paleontology: Worthless branch of science.
Parallax: Future perfect tense of *paralyzed.*
Parallel: Two lines, forever apart, never to merge, converge, or diverge; yearning, desiring, yet each alone.
Parity: Nickname for a city in France.
Parsec: A long way off.

Periodic Table: Where menstruating women have picnics.
Petri Dish: Casserole brought by scientists to potlucks.
Phase: An excuse for adolescent behavior.
Photography: The art of soul stealing.
Photosynthesis: A type of exorcism through which the subject of a photograph retrieves his or her soul.
Protein: An avid supporter of adolescents.
Quasar: Type of TV and light.
Radar: Whyearning, desiring, yet each alone.
Radio: Type of wave; type of entertainment; source of Dr. Science's fame and glory; well adapted to Dr. Science's mellifluous speaking voice.
Rayon: Miracle fiber.
Reflection: Wistful state of mind; what vampires can't see.
Refraction: Broken light.
Refrigeration: Artificial cold.
Resistance: What Old Science has to New Science.
Respiration: Sweat.
Retort: A snappy comeback.
Rocket: Improbable method of space travel.
Saccharine: Crude form of Nutrasweet.
Satellite: Sends messages to dishes.
Semiconductor: Part-time musician.
Series: Five baseball games.
Shadow: Popular radio character popularized, I believe, by Orson Welles.
Software: Hardware left out in the sun.
Solenoid: Painful condition developed by generators if they sit in one place too long.
Space: What fills the gap between objects.
Spectrum: Basic unit of saliva.
Square: Disdainful name for computer programmers.
Stainless: Mythical condition.
Stars: Distant bright objects; certain actors and television personalities.
Sunspots: Where photons go for vacation.
Symbol: Shorthand used to describe phenomena to nonscientists.

Synapse: Thought highway.
Syzygy: Famous Polish physicist.
Tachometer: Electric nail driver.
Telephone: Modern miracle; pleasant chirping replaces harsh ringing; digital display allows you to see last number dialed; automatic redial gives you one-touch convenience; computer tones replace old-fashioned rotary dial to give access to many discount long-distance services; cellular phones give us cord-free communication anywhere, anytime.
Telescope: Device used to manufacture "distance rays," which make "small" or "distant" objects "large" or "close."
Television: Device used to display satellite signals.
Telsa Coil: Mattress ingredient.
Therm: Humorous-sounding word.
Time-Lapse Photography: Overused technique in nature documentaries.
Topology: Study of spinning objects.
Transformer: Gobot.
Trigonometry: Outmoded form of mathematics.
Unicellular: Amoebic means of transportation.
Universe: Largest unit of space.
Use: Utilize.
Utilize: Use.
Vaccine: Woman's name used in many Chuck Berry songs.
Vacuum: Quiet place
Valency: Apartment for rent.
Vertex: Potential name for automobile model.
Virus: Photon from hell.
Viscosity: Property of fat liquids.
Vitamins: Small capsules that hitch rides on photons and are then harvested and distributed to pharmacies.
Volt: First name of the creator of Disneyland.
Vulgar Fraction: The so-called "number from hell." To write this number down is to invite eternal damnation; to use to in an equation is to invite the end of the world.

Wave: How light travels; how radio travels; how micros travel; what surfers ride; what we do when our relatives leave; how nausea travels; women's service branch of the Navy.

Work: How we pay for fun.

X ray: Convenient way to measure one's shoe size.

Yeast: Small creature who gives its life to give us bread.

Zenith: The most distant point in orbit; type of television; popular woman's name.

Zero: What one cannot divide by.

Zoology: Study of cages.

Appendix F

Bibliography

Our Friend the Atom, Walt Disney, Mouse Press, 1958.

How to Build a Cyclotron Out of Your Water Heater and a Few Spare Parts, Buster Wortman, Hobbyist Press, 1966.

Flying Saucers Are Real!, Dennis Ketcham, Nonsense Publication, 1963.

Cesium, Your Best Friend, Robert Lonely, Nostradamus Press, 1985.

1001 Things You Can Make Out of Radioactive Waste, H. R. Oppenheimer, Pegasus and Orion, 1944.

Radio Ranch, What Was Gene Autry Really Trying to Say? Gabby Hayes, Trigger Publications, 1951.

Worlds in Collision, I. Velikovsky. Comet Press, 1956.

The Tree I Had, Rev. "Bob" Davidson, DDS. Grove Press, Ltd. 1981.

It's a Wonderful Life! Frank Capra, 1948.

The Bicycle Thief, Vittorio De Sica, 1948.

East of Eden, Elia Kazan, 1955

Amarcord, Frederico Fellini, 1975.

Neatness Counts! Iowa Film Education Board, 1955.

Courtesy Pays and How! Iowa Film Education Board, 1954.

Proper Lunchroom Etiquette, Wisconsin Educational Film Council, 1956.

Mondo Cane, GrotesqueMex Films, 1968.

Anything with Doug McClure.

Appendix G

About Dr. Science Questions

Caution! Questions used in this book and on the radio program are absolutely guaranteed to be either authentic or fabricated. In fact, you are cordially invited to "Ask Dr. Science" yourself. If chosen, your question could appear on the radio show, in a subsequent book, or (depending on sales) in a photocopied pamphlet distributed free at Laundromats everywhere. You may also wish to simply convey your name, address, and zip code to us in return for a highly respected position on the Dr. Science/Duck's Breath mailing list. A good, healthy, and invigorating time for all is guaranteed you and your inquiries, scientific or otherwise:

Dr. Science/Duck's Breath
PO Box 22513
San Francisco, CA 94122

Appendix H

About Duck's Breath

Duck's Breath Mystery Theatre was founded in Iowa City in 1975 when five lonely, overeducated, and soon-to-be-underemployed types found a special sort of magic when they yelled at each other a lot onstage. Since then, the troupe—with original members Leon Martell, Jim Turner, Bill Allard, Coffey and Kessler still in tow—has bounded from coast to shining coast via live performances, radio, TV, print, tape, frequent flyer programs, and their own self-described "spunk." Now based in San Francisco, they have authored *Zarda: Cow from Hell!*, surely the greatest movie never made, and produced radio stuff like "Homemade Radio," "Ask Dr. Science," and innumerable characters and sketches for "All Things Considered." Collections of radio work are available on Newman Tapes. Television credits include staff writers for the Nickelodeon kids cable series, "Out of Control." Their most recent album, *Born to Be Tiled*, is available on Rounder Records. Since achieving international stardom, they have refused all interviews and awards, except when asked politely.

Appendix 9

Dr. Science's Hope for Mankind

It is my hope that Man will soon learn to control the gravitational forces. Once this has been accomplished, hunger and unemployment will vanish. Poor people will simply be able to float to a happier world, where there are food and jobs.

I also dream that someday, someone will erect a monument to algebra. I'd do it myself but my schedule is too full. The logarithm, the parabola, the hyperbola, differential equations—each aspect would be represented, symbolically.

I see a monument of polished onyx, set in a titanium foundation, somewhere in downtown East St. Louis, Illinois. The symbolism of this is entirely personal. It was in East St. Louis that I first learned the mysteries and delights of algebra. That was a long time ago, and East St. Louis, as well as algebra, has changed since then. I wish I could say they have both changed for the better but, alas, I am bound to tell only the truth.